Cosmologic Triadic Drive Theory:
The Physics of the Psyche

William L. Johnson

Consilience Series

Cosmologic Triadic Drive Theory: The Physics of the Psyche

By William L. Johnson, PhD

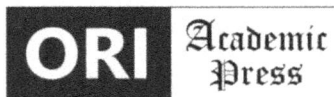

ORI Academic Press

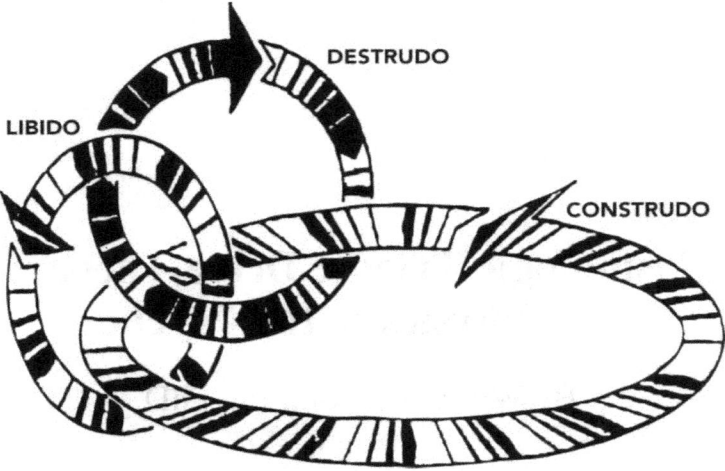

Published in 2018 by the
ORI Academic Press, New York, NY

Printed in the United States of America on acid free paper.

Library of Congress Control Number: 2018933837

Cataloging Data:

Johnson, William L. Cosmologic Triadic Drive Theory: The Physics of the
Psyche / William L. Johnson. Consilience Series.

1. Psychology and Physics – theoretical aspects. 2. Psychology and
Cosmology – theoretical aspects. 3. Dynamic Unconscious – theoretical
aspects. 4. Theory of Mind. 5. Triadic Drive Theory – psychological
aspects.6. Triadic Drive Theory – clinical applications.

ISBN-13: 978-1-942431-08-4 (soft cover)

mindmendmedia
piecing it together

Book design, editing, and book cover - by MindMendMedia, Inc. @
MindMendMedia.com

Dedicated posthumously to the author,
Dr. William L. Johnson
(1932-2014)

ABOUT THE AUTHOR

The author of this book, Dr. William L. Johnson (1932-2014) was a psychologist and psychoanalyst. He received a B.A. in Psychology from Queens College, a Master's Degree in Psychology from The New School for Social Research, an M.S.Ed in Clinical Psychology from City College of New York, and a PhD in Clinical Psychology from Yeshiva University. He later earned a post-graduate certificate in psychoanalysis from Adelphi University.

During his psychoanalytic training, Dr. Johnson studied with one of Sigmund Freud's first students, Theodor Reik. In the 80s, Dr. Johnson was an associate professor in the doctoral programs in clinical psychology at both Adelphi University and New York University. After being appointed chief psychologist at Orange County Mental Health Clinic in Goshen, NY, in the early 1960s, Dr. Johnson began private practice in Monroe, NY and Greenwich Village, NY City. He maintained both for nearly 50 years until his death.

Theodor Reik and William L. Johnson

TABLE OF CONTENTS

FOREWORD xv

PUBLISHER'S FOREWORD xvii

PREFACE xix

ACKNOWLEDGMENTS xxi

CHAPTER 1: INTRODUCTION TO TRIADIC DRIVE THEORY

The Universal Forces 3

The Constructive Universal "Building"
Forces of Physics..5

The Destructive Universal "Breaking Down"
Forces of Physics 8

Libido..15

CHAPTER 2: UNIVERSAL FORCES OF PHYSICS AND BIOLOGY

The Universal Alternating Influences 17

Psychoanalytic Principles Supported as Forces of Physics 22

The Unconscious 26

The Drive Ideal 32

CHAPTER 3: THE ORIGINS OF THE CONSTRUCTIVE DRIVE

The Combining, Connecting, Constructing, Creative Drive:
Construdo 35

Introducing a Third Drive 37

The Construdo Drive's Creative Level in Object Relations 38

The Origins of Construdo 38

Developmental Stages of Construdo 39

CHAPTER 4: DEVELOPMENTAL STAGES OF CONSTRUDO

Connecting and Disconnecting with Self or Objects:
Construdo Stages and their Derivatives 45

The First Intrauterine Stage: Self-Combining 45

The Self-Connecting and Self-Constructing Stage 46

Self-Creating Stage 47

Construdo Is Behind the Energy of Cathexis 49

Returning to the Fetus's Unconscious Development of
Construdo 50

The Newborn and the Combining Stage 52

Construdo Leads the Infant to Reality................................ 54

The Intrapsychic and Interpersonal Factors in the
Connecting Stages of Construdo (6 mo to 1 ½ years) 55

Interpersonal and Intrapsychic Influences of the
Constructing Stage of Construdo (1½ to 3 years) 57

Interpersonal and Intrapsychic Influences of the
Creative stage of Construdo (3½ to 6 ½ years) 58

CHAPTER 5: CONCEPTS OF THE DESTRUCTIVE DRIVE

Psychologies and Theories of the Destructive Drive
Contrasted with Destrudo 65

CHAPTER 6: THE SEVEN DEVELOPMENTAL STAGES OF SELF
AND OBJECT DESTRUDO

The Concept of Destrudo Development 71

The Self as Object 75

Destrudo in the Oedipal Conflict……………………..…… 83

CHAPTER 7: AROUSING DESTRUDO OR CONSTRUDO

The Arousal of Object-Destrudo…………………………...85

The Pathway from Object-Destrudo Arousal to
Self-Destrudo Arousal……………………………….…. 86

Causes of Self-Destrudo ………………………………… 86

Early Family Influences on the Arousal of
Self-Destrudo……………………………………………....87

Arousal of Object Construdo……………………………95

Arousal of Object-Combining Construdo…………………96

Arousal of Self-Construdo ……………………………… ...98

CHAPTER 8: FREUD'S DUAL DRIVE THEORY CONTRASTED
WITH TRIADIC DRIVE THEORY …………………………………101

The Theory of Three Drives …………………………………102

The Drives and Physics …………………………………...108

CHAPTER 9: THE UNCONSCIOUS AND THE SUBATOMIC
BRAIN FIELD: FREUD'S DREAM DISCOVERIES
OF THE UNCONSCIOUS

Creation of the Unconscious by the Subatomic Brain
Field and the Influence of Construdo in Dream
Formation …………………………………………… 111

Freud's Timeless Spaceless Dynamic Unconscious and the
Interior of the Atom……………………………………..... 115

Distortions of the Dream Work and the Physics of the
Dynamic Operations of Subatomic Particles within
the Brain's Atoms…………………………………….....119

Time and Physics …………………………………....124

Representation through Symbolism 125

Time, Space, and Freud's Unconscious 127

Freud Unconsciously Intuits Construdo 128

CHAPTER 10: FREUD'S "DREAM WORK" AND
UNCONSCIOUS CONSTRUDO

"False Connections" of Construdo Reasoning and
Freud's Secondary Revision of Dreams 129

CHAPTER 11: PSYCHOPATHOLOGY

Transformed Id Drives Distort Ego Function in
Space-Time... 133

CHAPTER 12: DRIVE PATHOLOGY AND HEALTH:
THE SPACE-TIME DIMENSION

The Dimension of Space-Time 139

Why Space and Time Are Not Absolutes 140

Psychopathology Caused by the Influences of Three
Discordant Realities on Construdo and Destrudo in
One's Life Events in Space-Time 143

The Time Factor in Freud's Unconscious Dream Work..........146

CHAPTER 13: BIOLOGICAL LIBIDO DRIVE CONTRASTED
WITH DRIVES DERIVED FROM PHYSICS

Introduction 153

Feelings of Alienation... 155

A Case Example 155

Freud's Concepts of the Instincts or Drive 158

Neurophysiological Data on Nerve Transmission 160

Returning to Triadic Drive Theory.....................................164

Freud's Observation of the Vicissitudes and Outcomes of
Drives 166

CHAPTER 14: PARAMETERS OF DREAMS AND REALITY

The Dream Parameters 171

Reality Parameters of Construdo and Destrudo 178

CHAPTER 15: PSYCHOPATHOLOGY IN THE DRIVES

Imbalances of Forces from Our Universe, from Our
Planet, from Our Brains' Atoms Cause Drive Pathology 181

Freud's Concept of a Constant Drive Influence 186

Indications in Dreams of Time Delays in
Unconscious Operations 186

Other Realities Causing Drive Psychopathology 187

Drive Reversals and Other Realities 188

CHAPTER 16: THE CONCEPT OF CHARACTER TRAITS

Freud's Concept of Character Trait Formation and
TDT's Additions 195

Retreat, Repression and Regression 197

The Space-Time Factor in Drive Conflicts 199

Confusion in Couples about Their Drives 200

CHAPTER 17: FORMATION OF DESTRUDO
CHARACTER TRAITS

Destrudo-Derived Traits 203

CHAPTER 18: FORMATION OF CONSTRUDO
CHARACTER TRAITS

Construdo Traits Derived in Utero 215

Construdo Traits Derived after Birth .... .219

CHAPTER 19: THE DRIVE IDEAL

Construdo Drive Ideal Expressions and Combinations 231

A Restricted Drive Ideal 234

The Power of the Drive Ideal .... .234

Destrudo Influences on the Drive Ideal 235

CHAPTER 20: TDT'S CONTRIBUTION TO HUMAN
SELF-UNDERSTANDING: 17 ASSUMPTIONS
AND 16 SPECULATIONS, A SUMMARY

The Assumptions of TDT 239

The Speculations of TDT 243

Regulating the Human Drives: The Drive Ideal 251

AFTERWORD 255

REFERENCES 257

GLOSSARY 261

FIGURES AND TABLES:

Figure 1. Triadic drive energy interactions 109

Figure 2. The space-time dimension in modern physics........... 151

FOREWORD

Professor Johnson has creatively synthesized in this bold, break-through volume Freud's long neglected metapsychology regarding the dyadic drive theory (libido, death wish). The death wish hypotheses have always been deemed the most speculative and unscientific aspects of Freud's theoretical writings. The idea of an aggressive drive seemed to continuously run counter to more behavioral observations linked to a frustration-aggression hypothesis which had been seemingly more comprehensively studied and empirically verified.

Professor Johnson's most creative contributions, in this reviewer's opinion, are his shift to a triadic drive theory (libido or love, construdo and destrudo). The latter two drive forces stem from and parallel the natural physics based energies and forces of the universe rather than those biology based forces favored by Freud. Those energies that pull together a multitude of entities are linked to Construdo, and those that violently break things apart – to Destrudo. The Big Bang theory of the origins of the universe as we know it can be conceptually linked to both of these forces.

This brilliant, courageously developed volume shows a freedom of mind rarely seen in contemporary psychoanalytic theories which tend to neglect the more energy based, drive aspects of human behavior. Even more so, those destructive and self-destructive behaviors are linked to the frustration-aggression hypothesis and hence are seen as learned rather than instinctive forces.

How can we explain the universal destructive repetitions on the macro level in wars (currently via ISIS; historically, the Pearl Harbor and culminating in Hiroshima and Nagasaki nuclear bombings; the Holocaust; and other examples of wild genocidal and suicidal bombing attacks against vulnerable human beings) and on the micro clinical level of self destructive attacks of political figures (like Elliot Spitzer and Anthony Wiener) against themselves and their powerful attainments – without hypothesizing a physics based destructive energy?

Professor Johnson has written a very modern theoretical volume that has vast practical clinical and social implications for our contemporary world. I admire very much his willingness to grapple with and update rather brilliantly Freud's much neglected drive theories.

Morton Kissen, Ph.D.
Professor Emeritus, Adelphi University

PUBLISHER'S FOREWORD

> We cannot do without people who have the courage to
> think something new before they can demonstrate it.
> (Freud to Fliess letter, 1895)

How is our mentality created? How does the human mind develop? Are there any natural or super-natural forces present through one's life that shape this person's (material) brain and (non-material) mind/ psyche? Are the material brain and non-material mind connected in some way, and do they change each other as they operate, even from before birth? Are both the brain and the mind of one person connected in any way to the minds and brains of others? All these are the fundamental questions that great people of the 19th and 2oth centuries explored, and now, in the 21st century, their followers continue this endeavor. So, this volume, by a brilliant psychoanalyst Dr. William L. Johnson, offers some answers to the above questions, examining them through the lens of the Triadic Drive Theory (TDT), which explains how the forces of the universe create our mentality. Triadic Drive Theory is built on Freud's dual drive theory, which relies mostly on biological concepts, but in this work, Dr. Johnson uses the laws and the language of physics, cosmos and the universe to explore one's mental life and one's relationships.

At the dawn of development of psychoanalytic theory, Sigmund Freud was working on his *Project for a Scientific Psychology*, which had the intention to "furnish a psychology that shall be a natural science; that is, to represent psychical processes as quantitatively determinate state of specifiable material particles," as Freud described it himself. Most of Freud's concepts, – e.g. unconscious and conscious; ego, id, and superego; primary process and pleasure principle; secondary process and reality principle – trace back to this *Project*, which was soon abandoned by him – most likely because Freud was marching ahead of his time, ahead of scientific developments, especially developments in neuroscience, and he could not afford to jeopardize the validity of his creation, psychoanalysis, by not being able to explain any of its conceptualizations using the scientific method.

After reading and re-reading the manuscript, one cannot hold off the feeling of awe regarding Dr. Johnson's work in general, and especially his courage to express the "out-of-the-box" ideas, to question the status quo of Cartesian dualism, and to allow the unity of knowledge – as in Edward O. Wilson's (1998) principle of consilience – to shine. Sigmund Freud comes to mind here, as he insisted that one should not be "dissuaded from

xvii

publicly expressing [his/her] views, even if they are only conjectures" (Freud to Fliess letter, 1895). In the meantime, Dr. Johnson presents here not just the "conjectures," but a very clear drive theory based on his knowledge and practice of psychoanalysis and his comprehensive research related to physical processes, cosmos, and the universe, which led him to seventeen assumptions and sixteen speculations that formed the foundation of his theory.

First, TDT assumes that all mental processes derive from "material" operations of one's brain, even brain particles (something that Dr. Eric Kandel, a neuroscientist, proposed in his "New Intellectual Framework for Psychiatry" in 1998, and was crucified for). Second, TDT assumes that Freud's biologic drive, libido (the human sensual, sexual drive) operates in conjunction with two drives that derive from physics – construdo and destrudo; and this way, uniting the approaches of two basic sciences, biology and physics, in one. Third, interdisciplinarian Johnson unites various grand ideas, like the Big Bang and Big Church phenomena offered by Stephen Hawking; quantum mechanics of Werner Heisenberg; Eastern mysticism and physics of Fritjof Capra; psycho-socio-historical analysis of Erich Fromm; Sigmind Freud's understanding of dreams, unconscious, biological drives, and creativity; as well as the forever concepts of religion and God; good-versus-evil, and many more – into one (triadic drive) theory, and applies it to all possible aspects of human mentality.

We are sure that those who read this volume with their minds open to unconventional ideas will find a lot of answers to some very fundamental questions related to what it means to be human. And, we hope that Dr. Johnson's work will become an inspiration for others to uphold the unity of the knowledge principle, to "think something new before they can demonstrate it," and to find the courage to bring their ideas for discussion in a public forum. We are honored that Dr. Johnson's loved ones trusted us with bringing this book to life. Enjoy!

On behalf of ORI Academic Press,
Dr. Inna Rozentsvit

PREFACE

Sigmund Freud left us a "dual drive" theory. He added the destructive or aggressive drive in 1920 to the sexual and procreating libido drive, which he had presented in 1905 (Freud, 1905/1953b; 1920/1955). He made the assumption that the destructive drive, like the libido drive, was biological. Psychoanalysis has followed this assumption. Biology, however, has never explained the origin of the drive.

I began to wonder if the reason biology did not confirm the origin of the destructive drive might be because the destructive drive was not, in fact, biological. And if the destructive drive was not biological, what could it be?

I thought about stages of aggressive development that were related to the musculature and its development, and the muscular capacities accompanying this development. In 1920, Freud said that the musculature was the executor of the aggressive drive (Freud, 1920/1955a). As I continued to think about it, it occurred to me that when the subatomic particles of the brain smash into each other, it could actually be the origin of the aggressive (destructive) drive. From this smashing event, new subatomic particles are created. I realized that this phenomenon could represent a change in "origin" theory from biology to physics.

Once this fundamental shift in theory of the origins of the destructive drive from biology to physics was made, it followed that physics might be the basis for other observable human behaviors.

I was unsure of where this shift in thinking would lead me, but I was eager to explore the connection between human beings and the fundamental forces of the universe.

This book is the result of that exploration.

ACKNOWLEDGMENTS

The idea for writing this book began years ago when my friend Dr. Henry Kellerman asked me to write a book for a series he was editing.

I listened to his assignment regarding the subject matter of the proposed book but declined, telling him I was more involved with a different topic – one I had been thinking about for some time.

He listened, it interested him, and he agreed to take it on. His endorsement was encouragement to continue with this work, and I will be forever grateful.

I would also like to thank Helen McCabe, Rowena McDade, Charles Neighbors, Penelope Hull, Nancy Monichelli, and Pamela Byrd for their support.

Note: Fortunately, Dr. Johnson completed this work before he died. The final edit was addressed by his colleague and friend, Dr. Henry Kellerman, and supported with great and loving efforts by his longtime partner, Pamela Byrd, and his daughters, Ms. Toni Ann Johnson and Dr. Hillary Johnson.

CHAPTER 1

INTRODUCTION TO TRIADIC DRIVE THEORY

Triadic Drive Theory (TDT) explains how the forces of the universe create our mentality. It begins from a different reality about the nature of human beings than has been commonly held by psychology and psychoanalysis. TDT's initial proposition is that the constant forces of physics throughout the universe and in the operations of subatomic particles in the atoms of our brain cells produce the energy of our two most basic life-determining psychic drives.

Construdo is the drive that causes us to combine and connect thoughts, ideas, emotions, concepts, attitudes, and beliefs, as well as to connect with other cultures, societies, families, and people. The second drive, *destrudo*, determines why, when, and how such unions are broken apart, damaged, or overthrown. These two drives and their derivatives, it is proposed, determine more human behaviors than Freud's libido drive and its derivatives. TDT builds on Freud's dual-drive (libido and aggressive or destructive) theory. However, it rejects or refines Freud's assumption that both the libido drive and the destructive drive are biological.

TDT begins with the known reality of the universe. Human beings exist on a planet that is a single minute speck in a galaxy of more than one hundred billion stars. The sun in Earth's solar system is but one medium-sized star in the Milky Way galaxy; there are about a hundred billion galaxies in the universe, each containing about the same number of stars. Considered by itself, our solar system is a microcosm in the universe.

We take life for granted on our planet, but in the rest of the universe (as far as we know) life is exceedingly rare, if not nonexistent or perhaps more accurately, undetectable.

Triadic Drive Theory asserts that the forces of physics that govern and hold the universe together constitute the true universal reality for humans as well as for all other species. Thus we can propose an axiom: we are all governed by the same principles of physics that govern the rest of the universe. How these forces of physics produce our cognitive or mental drives is the focus here of theoretical, psychoanalytic, and behavioral observations that have led to the formulation of Triadic Drive Theory.

THE UNIVERSAL FORCES

We know of four forces in the universe: gravity, electromagnetic force, weak force, and strong force. In the earliest instance of the universe,

3

theory has it that there was a tightly compacted energy particle held together by binding, uniting forces of physics, which exploded in the Big Bang, beginning the universe. The uniting, binding forces on matter then continued as separate, differentiating forces in the expanding universe that followed the Big Bang.

After the Big Bang, gravity pulled the expanding dust, gas, and dark matter back toward each other and then later formed universal bodies in opposition to the expanding energy from the Big Bang. When gravity finally overcame expanding energy, it pulled passive universal bodies back into each other in a contraction. That process, according to physicist Stephen Hawking (1988), would be the "Big Crunch." Hawking said that the Big Crunch is only one of three possibilities, but it is the possibility that TDT supports, as it is consistent with observations that could be predicted by TDT.

The electromagnetic force in the universe holds atoms and molecules together. If not for this force, we could walk through walls. The electromagnetic force operating between particles prevents this, because binding forces between the particles that make up a person and a wall block such passage. The molecules of all animate and inanimate objects are held together by electromagnetism, as are parts of atoms that comprise them.

In addition, within molecules, the electrons of atoms revolve around a nucleus held together by electromagnetism. This weak nuclear force also causes radioactive decay of the atom's nucleus.

The nucleus of all atoms, including those in our brains, is made up of protons and neutrons bound together by the strong nuclear force. Other binding forces may be yet be discovered, but that does not affect Triadic Drive Theory; these forces would simply be other binding forces. These binding forces underlie a psychic uniting and combining drive. These four binding forces of physics: gravity, electromagnetic force, weak nuclear force and strong nuclear force hold together all that we know of as reality throughout the universe of living and nonliving things. It is proposed that these four forces also cause a psychic drive to crystallize within our brains and, as a hypothetical construct, into our psyches. Thus, what exists in our bodies and brain cells has such so-called psychic representations.

TDT posits that a system's overall operations influence its inner parts to operate similarly. Therefore, it may be assumed that the way the universe operates as a whole system is reflected in the way we, as one of its infinitesimal parts, operate. This influence begins in the fetal brain during the second and third trimesters, producing the fetal brain's theoretically proposed unconscious psyche. Of course, fetal body formation influences the fetal unconscious psyche with an energy that binds body cells into tissues, organs, and organ systems, following the proposition that physiological functions are intricately linked and interchangeable with psychic functions at the unconscious level. Therefore, it is proposed that the way the biological body operates reflects the way the human psyche operates. In other words, the psyche reflects both cellular and physiological processes.

Moreover, cell proximity and similarity (Gestalt psychology grouping together factors and genetic proclivities) causes differentiation into tissues, organs, and organ systems, promoting larger and larger cellular groupings. Life itself develops under the influence of these forces. Interestingly, Freud realized that his dual-drive theory lacked a precise accounting of the force that caused such combining into larger and larger cellular groupings.

THE CONSTRUCTIVE, UNIVERSAL "BUILDING" FORCES OF PHYSICS

Binding and uniting forces converge on the fetal brain to produce a psychic drive that combines and connects with whatever the fetus senses around it within the uterus. It's as though the fetal brain can actually sense the mother's positive and negative attitudes toward its existence. Met with imagination, it can be considered positive that these attitudes either become part of the fetus's deep reactions to itself or are intensely reacted against. The way the fetus has reacted possibly – again, with suspended disbelief – can be deduced from personality tendencies in the adult.

During the second trimester, TDT proposes that the unconscious psyche of the fetus can combine and connect itself with its uterine surroundings. The fetus is in a warm, fluid environment without light or sound. The tactile sensing of its body's fetal folds forms a part of the uterine environment. By the end of the second trimester, the fetal body has completely formed.

As the fetus enters the third trimester, it gains mass. Then, to continue the triadic drive theoretical position, the achieved self-combining and self-connecting aims yield to the emergence of a process of self-constructing – the fetus constructing itself. This sense of self-construction consolidates as an aim of the psyche during the third trimester. Thus, shortly before birth and at birth, the fetus has a sense of having created itself – the fourth and final self-uniting aim of the construdo drive.

Construdo can be seen to have completed four self-developmental stages: self-combining, self-connecting, self-constructing, and finally a self-creating stage. After birth, the four binding forces of physics that produced construdo continue. Now, however, there is a dual focus: first to combine, connect, construct, and create the self, and second to do the same with the surroundings. Patterns constructed from earlier combinations and connec-

5

tions with the surroundings lead to new constructions and new perceptions we can identity as new creations in the environment. Visual images in utero and thoughts, reactions, concepts, and emotional responses are all subject to the overall triadic drive's influence to connect with and construct things following birth. This drive, which begins in utero, ultimately produces, it is predicted, the fetal unconscious.

Disconnecting Construdo

In the third trimester, drive energies begin to differentiate or disconnect from each other. Disconnecting from construdo pursuits in the unconscious psyche means striking those connections out of the unconscious psyche or pushing them out of their construdo position – in other words, not allowing the pursuit any time in the psyche of construdo to gain the ascendancy over differentiated drive energies or any ability to achieve connection.

This destrudo effect is accomplished by the differentiating reactions of the destructive drive's energies. The disconnecting process continues after birth and then throughout life, because the two psychic energies (construdo and destrudo) are different. They are antagonistic, don't fit together, and have different directions in time. As differentiation triumphs, of course, these two drive energies separate further from each other.

Yet, in the sixth month after birth, there is a transitional construdo stage from the original self-constructing aims in utero, which was for the fetus to combine and connect with the new external environment. It needs to be remembered that combining is construdo's first and most fundamental goal.

From six months to one and a half years, an infant's construdo drive leads it to an interactive connection with its surroundings and with itself in its new, post-uterine environment. According to Jean Piaget (1954), the infant rejects the constantly changing perceptions of its unconscious. Piaget called this "undifferentiated chaos." This changeability of perceptions is the way the unconscious perceives reality. The infant wants to disconnect (or reject) reality because it is so unpredictable and mentally disconcerting. Thus, unconscious perceptions change shapes. Perceptions reverse themselves. Objects get larger, then smaller. Objects are upside down, then right side up. The infant seeks to halt these confusing perceptions. Freud (1915/1963) saw this goal of "halting" as the beginning of the "reality principle." It is when the infant strives to locate reality within the environment with which it is trying to connect.

A year and a half and later, the child is constructing new thoughts and feelings about the world and continually constructs things in its play. The child mentally constructs an image of what it wants to be. The child thus creates itself beyond self-invention. The self-construction and object-constructing pursuits continue until about age three and a half. After that, the child combines, connects, and constructs from past images, perceptions, thoughts, and feelings, which allow the child to create itself in unique and creative ways and to develop into the being it wants to become via the unconscious. These creations have profound appeal, because they are manufactured by unconscious processes; hence, such creations reflect a fusion of thinking and the deeper feelings that exist in the human unconscious. This fourth and final stage – the creative stage – takes place from about age three and a half to six and a half.

When individuals become overly invested in any construdo stage-aim, the focus to return again and again to such underlying energy origins can lead to the formation of construdo character traits. These are permanently etched personality traits that identify the person. For example, people who possess an elevated need to combine with others – overly connected individuals – are considered to demonstrate a clinging trait. Overly constructed individuals are seen as workaholics. Overly creative individuals are deemed artistic, and so forth.

Construdo Arousal

The construdo drive is aroused when distinct unrelated stimuli, from people or things, confront us. We try to understand the stimuli, first by combining and connecting them (or the event) with something we understand from the past, likening the combined connection of the event to our present understanding of our current surroundings. The concept we use to make this connection may sometimes be erroneous, but it permits individuals to believe they have validly connected with events, people, or things, even though it may be a "false connection," as Freud (1912/1958a) called it. Freud's term refers to connections by the unconscious carried into consciousness. TDT concurs in the understanding of false connection, but considers it as being "overly connected."

Construdo Love

Love is a creative stage derivative of construdo. In love we create an ideal image of the loved person. Love is a construdo creation of the psyche. Reality testing is suspended in love, as in when we see a loved one through rose-colored glasses. Lovers may share fantasies of how their love will cause them to act with each other; the more similar those developing fantasies, the more they feel deeply in love. Hence, most mentally healthy people reach the creative stage of construdo when in love. Psychotics are self-creative when they create delusions of themselves.

Construdo's persistent, unconscious influences can be observed when we consider a behavior such as greed. First, individuals seek to combine by acquiring more money, land, power, knowledge, possessions – whatever is more than others have. They become aggressively competitive, and this aggression further fuels their acquisitiveness. But once things have been acquired, the person is driven toward more combining. Seeking to combine further is conscious, but the underlying influence of construdo acquisitiveness and competitive aggression is not conscious. Addictions work similarly. Once a level of substance indulgence has been established, the individual craves more, because construdo continues to seek new combining with more and more of the substance. Construdo drives individuals into the level of addictive need.

The Life Drive

Construdo's constant influence on humans exists, though individuals have no conscious awareness of it. Sigmund Freud (1915/1957b) described this as the essence of the instinctive psychic drive. Construdo is the life drive. It furthers "building up" processes at all levels of our existence.

Combating Illness

It can be predicted by TDT, therefore, that reinforcing construdo bolsters our immune system, thus combating disease in the body. When illness arouses self-aggressive impulses, one's self-construdo impulses can counteract them. Such a counter-reaction can create a mental framework for recovering health by superseding the destructive drive of the illness process as one reinforces construdo energies.

THE DESTRUCTIVE UNIVERSAL "BREAKING DOWN" FORCES OF PHYSICS

The second drive of the psyche considered by the Triadic Drive Theory is understood to be generated by universal forces of physics, and it hypothetically emerges from the larger destructive forces occurring throughout the universe, producing a comparable destructive drive in humans. This connection of the macro to the micro ostensibly began in the theorized Big Bang explosion of the original, tightly compacted, super-dense energy particle. As indicated earlier, the Big Bang's resultant gases and later dust clouds cooled into matter that joined together and continued to expand outward as galaxies, stars, planets, satellites of planets, and dark matter. Such joining together of particles reflects the universal energy underpinning the proposed construdo drive.

Destructive energy tendencies are observed in galaxies colliding into each other, in stars exploding into supernovas, in stars smashing into each other, in comets smashing into satellites and planets, and in asteroids and meteorites smashing into planets during the 13.8 billion year history of the universe.

Gravity energies are involved in the formation of black holes, where a star collapses back into itself by the force of gravity. Anything coming within the energy influence of a black hole is drawn into it, and it is so compressed by gravity forces that presumably it can never exist again. Gravity operating in this destructive way is stored as potential destructive energy. Again, in a microcosmic enactment of the macrocosmic model, it may be considered that this storing of potential destructive energy is a second contributing factor to the destructive drive in humankind's unconscious.

The potential energy in universal bodies that can cause explosive reactions when they smash into each other, in a jump to a human model, would be seen as one that energizes the human fetal unconscious with equivalent potential energy enabling a similar reaction. When these universal bodies collide with each other, energy is released in all directions without specific focus or aim. It is a prototype of undirected destructive energy release on the human level. The influence of such potential energy release would therefore be implanted in the cells, molecules, and atoms of the forming fetal brain. This potential energy would be a source of influence on humans – that is, the universe's operations, as a total energy system, creating a destructive drive within us. Thus Triadic Drive Theory assumes that the whole universe and its operations exert determining influences on how its microscopic parts (each of us) will operate. All of the parts will follow the organization of the energy systems of the whole universe.

On Earth there are many natural destructive energy influences on our unconscious: for example, volcanic eruptions, fires, tornadoes, cyclones, typhoons, hurricanes, tidal waves, and earthquakes. These destructive forces of nature unleash tremendously destructive energies on our planet. TDT thus proposes that when we live in proximity to these events or hear of them, they directly influence our own destructive unconscious. Thus, it may be that even when we are not physically close to these destructive energies, they are learned models for potential types of destructive energy releases in the human unconscious.

Parental Influences

The way our parents express anger or show restraint is a most important, expressive influence on the formation of destrudo drive expressions and how we will or will not display it. We may conversely express behaviors in opposition to our parents, depending on whether we approve or disapprove of their destrudo expression.

Subatomic Particle Creation

The electromagnetic force binds the electrons to the atom's nucleus. The weak nuclear force causes radioactive decay of the atom's nucleus, a slow destructive process. As an analogy to the dynamics in humans, the weak nuclear force can be seen as a crossover in its energy from constructive energy sources to destructive energy sources in the brain's atoms.

The annihilation of particles and the creation of new particles can also be seen as a crossover at the subatomic (unconscious) level to creative construdo aims. Out of these destructive energy encounters, new particles or new solutions are added to creative problems held in the unconscious. Creation involves drive energy reversal of aims in consciousness by the unconscious – from destruction to creation, back to destruction, and so on.

Destrudo

The phenomenon of subatomic brain particles smashing into each other (combining) and then annihilating each other is another basis for understanding unconscious destructiveness in human beings. Explosive energy came into the universe initially during the Big Bang explosion that created the universe. This universal influence continues as galaxies collide, as stars smash into each other within galaxies, as comets, asteroids, and meteoroids smash into other celestial bodies, and in the furies of nature on earth. All of these explosive events demonstrate universal energies that can destroy what exists. A basis of human destructiveness in the context of our sociology is, it is proposed, derived from these destructive events and their registered influence on fetal unconsciousness.

These phenomena of destructiveness are examples in the fetal unconscious of how a psychic drive for destruction arises manifesting in a drive to destroy oneself, one's mental events, or things and people in one's surroundings. This destructive drive expands into consciousness following birth. An expanding construdo influence in part joins the destructive energies around an individual. Thus, this model even predicts that planetary furies will ultimately produce models of actions for individuals via their own actions in expressing destructiveness.

Within the body's cells there are anabolic processes (those that build up), and catabolic processes (those that break down). Building-up processes such as hydration synthesis and breaking-down processes such as dehydration synthesis occur within a cell's molecules. It is further proposed that these molecular processes contribute to the fetal unconscious's sensing of breaking-down processes (destrudo) and building up processes (construdo). Thus, an influence seems to exist to break down what surrounds us or what is inside us, contrasted with the construdo influences to build up things, or put things together. TDT (Triadic Drive Theory) calls this psychic drive destrudo, a term introduced into psychoanalysis by Edoardo Weiss (1960).

The unconscious in its microcosmic functioning is timeless in its receptivity to outside continuous influences from the universe. Destrudo's ultimate aims are backward in time, to destroy what has been established in the individual's environment or mind. Destrudo is a drive energy that seeks to remove things that block and frustrate the individual from reaching desired construdo goals, particularly those from back in time before any frustration existed. The destrudo drive is aroused when construdo antici-pations fail to materialize.

Destrudo's Developmental Stages

It is proposed that destrudo begins its developmental biological stages during the first six months of uterine existence when the self-destrudo aim is sensed by the embryo, and later by the fetus, as a potential difficulty, as in the threat of a miscarriage or spontaneous abortion. The self-destrudo phenomenon would break the embryo or fetus apart. In pregnancy, the practical causative factor could be a hormonal imbalance, a genetic defect, an external physical injury, or poor prenatal health. On rare occasions, the break-apart self-destrudo aim will show itself later in life. This is the first uterine stage aim of destrudo, a break apart phenomenon. Self-destrudo stages (destrudo directed against the self) begin at this point in utero, as does self-construdo. Self-directed stages of the two drives may be seen to develop when the fetus can sense only itself in utero.

Drives, both destrudo and construdo, consist of ultimate aims directed against the self in utero. Construdo combines and connects with the self, and it also links destrudo with the self in utero. These psychic drives, originating in our TDT (as a reference to the science of physics) influence the forming fetus in utero as soon as the brain has formed and can register such influences on the fetal psyche.

During the third trimester, the fetus may sense an organ out of balance with other organs and organ systems – an organ that may carry a genetic or other defect. A defective heart, lung, liver, or kidney may cause life to end soon after pregnancy begins, or the organ many improve so the newborn survives. In either event the fetus will have sensed the second self-destrudo aim, a defective part causing self-destrudo. TDT would ominously predict that this aim will recur later in life.

Birth initiates arousal of the newborn's object destrudo aim against the unexpected new surroundings; unrealized construdo anticipations of a continued uterine environment are suddenly disrupted by the newborn being delivered into a world outside of the uterus. Presumably, this is a shock for the newborn leaving its warm, fluid, visually unstimulating, quiet, odorless, nongustatory, tactilely comforting uterine environment.

TDT predicts that the confused newborn will react with intense negative destrudo toward the new environment; however, it has no obvious target to attack or retaliate against. While construdo's aim in space-time is to combine with the new surroundings, its energies also combine destrudo with the self. Even so, part of construdo will disconnect with these energies opposite to its own; hence destrudo will be both self-directed and object-directed.

Following birth, this process of the operation of destrudo and construdo in the newborn creates a transitional stage between uterine self-destrudo stages and object-destrudo stages, or aims toward the environment. This transitional stage marks the thrasher-destrudo stage, so named because of the newborn's muscular activity: kicking its feet and legs angrily and aimlessly, striking wildly with hands and arms while rapidly twisting its torso. The newborn vocalizes screams and cries in furious reaction to being delivered into a new, discomforting world. Again, this transitional stage initiates the vocal expression of destrudo.

The newborn's transitory self-aim is destructive to the self, seeking physical exhaustion during these angry energy expenditures. The newborn's beginning transitory object or environmental aim is to destroy whatever it encounters in the surrounding environment – the beginning of ambivalent, alternating construdo and destrudo aims. Simultaneously, the newborn wants both to join the environment and destroy it, but the newborn does not know how to distinguish between the two. So the baby is torn between both extremes (construdo and destrudo), a conflict that can arise throughout life whenever drive conflicts are fixated or drive development becomes arrested at such a space-time (the measurement in physics indicating a given event and when in time it occurred).

In time, the infant learns to visually and accurately perceive its surroundings. Between six months and one year, it develops depth perception and clearer vision of nearby objects, simultaneously developing a prehensile grasp (the ability to grip objects between thumb and forefinger), reflecting the infant's concomitant desire to master or control the objects it sees. When the infant can take hold of objects from two different directions, it gains a sense of control over or mastery of the object.

The essence of this stage's aim of destrudo is to catch the object in an attack from two opposite directions (pincer-object destrudo). It occurs between six months and one and a half years. Pincer-destrudo, being self-directed, stems from when the infant experiences the conflict or dilemma of being angry with its mother but dares not show it; the infant needs its mother to survive, but senses that its mother does not need it to survive. The mother could leave the infant and be fine. Thus it does not express anger toward her in fear of her departure. Consequently, the infant, with no choice toward whom to release the anger, releases such anger at itself when confronting a conflict.

From one and a half to two and a half years, the child acquires the ability to push or pull a toy to a desired location in relationship to itself. This new muscular power delights the child. It is expressed by repeatedly opening or closing a drawer or a cabinet door. The child translates this newly found ability into pushing or pulling a parent to a desired location in the environment when the child's connecting construdo wants the parent to see or do something at the desired location.

Now the child begins to reflect also on its relationship with people. Have they made it feel good or bad, to use Freud's terms; provided pleasure or "unpleasure"? If the child has been made to feel bad or unpleasant, then the child up to age two and a half wants to change or overturn the relation-ship with these people by moving them to some other place in relationship to itself. It's the toppler-object or toppler-self-destrudo aim. The object or self is toppled from a set position and then pushed or pulled into a desired new position in relation to oneself.

It is congruent with Freud's concept that the musculature is the executor of the aggressive drive. As the musculature acquires developmen-tal ability to perform different physical actions, the child begins to express different destrudo aims as its musculature progresses developmentally. It is the biological – physiological developmental component of the destrudo drive's unfolding in human beings.

By two and a half to four years of age, a child can kick an object and hit a target with its hand, a hammer, or a stick. It can throw a stone at an object as an extension of its hand's intention to hit the object. By this time the child has experienced pain both physically and emotionally, inflicted by parents, siblings, peers and others. These people have frustrated the child's desires to be treated well and not feel hurt by their behavior toward it. This hurt and pain arouses the child's destrudo, because its construdo anticipations are different and those anticipations are not met. At this age (two and a half to four years) the child begins to retaliate with similar destrudo responses. It wants to hurt and inflict pain upon individuals who have hurt or caused it pain. The child hits or kicks or throws something at the offender. Or, the child says hurtful things to those who have offended it. This is the *striker-destrudo* aim, a learned retaliatory response that occurs when the child perceives others to have inflicted hurt or pain first. It is executed in a sudden striking attack when and where the target individual is undefended.

Sometimes before striking activity, there is a *poking* activity – that is, physically touching, poking at the opponent's defenses to determine where an undefended place might be, like a jab in boxing, followed by a sudden hard punch, landing on an undefended spot. An adult might exhibit this poking activity in an argument wherein one tests the weakness in another's defense and then makes a striking verbal blow with a cutting remark.

It may also be contemplated that striker-self-destrudo is seen in a child's unwittingly running into a tree or even perhaps in an adult striking at the self by having a sudden heart attack. In a TDT context, either may result from unconscious guilt or the need for self-destrudo punishment.

The final destrudo aim occurs between ages five and six and a half. By then the child has jumped up and down on an object and smashed it into pieces. It has torn objects apart. It may have dismembered an insect. The child's actions may have caused a thing to no longer exist. The child's musculature is now capable of such actions and can express such destrudo aims in behavior when provoked or frustrated. But realistically the child can't direct such destrudo aims toward persons. However, the child may have fantasies of dealing similarly with people. Guilt over having these fantasies (particularly when the targets have been its parents) can, in clinical understanding, last a lifetime in a person's self-punishing uncon-scious.

Destroying an object so that it can never exist again is the final destrudo stage aim. It is the fifth object-destrudo aim after birth. It's the seventh self-destrudo aim considering uterine time aims. (Following birth, all destrudo aims have both self and object aims, or another person aims.) At this stage of destrudo the aim is called object-*smasher*-destrudo and self-

smasher-destrudo. One example of self-smasher-destrudo is a suicider who jumps off of a building and thus smashes the body to death.

Destrudo Arousal

Destrudo is aroused when a person's construdo expectations, plans, and fantasies are blocked by another or by events. Destrudo is also aroused when a person is hurt by another unexpectedly. When this happens, it results in retaliatory destrudo. When aroused to the smasher-object-destrudo aim (in a rage), some individuals become directed entirely by destrudo, seeking to smash all former construdo connections with the destrudo-arousing adversary.

LIBIDO

The libido drive was propounded by Freud as the sexual, procreating, species survival drive. It has been incorporated into TDT as the third drive. TDT accepts the oral, anal, phallic, and genital stage aims of Freud's libido drive.

TDT adds two uterine developmental stages to libido aims: first, the *self-forming* stage during the first six months in utero when the embryo and fetus are forming tissues, organs, and organ systems. Second, the *self-contactual* stage in which the folds of the fetus rub against one another, causing self-tactile sensations or self-sensual libido. At the same time, there are energies of concomitant construdo self-combining and self-connecting stages acting on the organism. Construdo energy aims here reinforce libido aims. The construdo self seeks contactual libidinal stimulation from the self.

Libido moves forward in space-time energies and development, because it combines with construdo. It also moves forward in space-time because growth is reinforced by construdo. Libido and construdo drives in tandem are in their most powerful and mentally influential arrangement in the person's development.

Following birth, this contactual need is seen in tandem with the newborn's need for construdo, combining and connecting; being held, touched, and stroked by caring adults. Children, adolescents, and adults generally continue to demonstrate this need. Its satisfaction reassures and reinforces the individual's sense of being loved. It occurs in construdo created love, in symbolic expression of individuals being lovingly touched and held.

Conversely, when object-destrudo aims are combined with libido aims, it results in some variety of sadism. Masochism results when the self-destrudo aims are combined with autoerotic libido aims. Sometimes an unconscious drive reversal causes sadism to turn into the conscious display of masochism. At drive extremes, we find the unconscious influence of alternation or reversal of the drive aim.

Libido Arousal

When someone with charming physical characteristics and impressive deportment is seen, in the mind's eye one may also perceive formidable spiritual attributes the person possesses. This person has certain qualities of voice, a certain smell, and if touched, certain feel. The different aspects are combined creatively in the beholder's mind, stimulating exciting libido fantasies of a construdo union. Such fantasy activity arouses libido. When aspects of these fantasies parallel earlier desired libido fantasies – conscious or unconscious – the libido arousal will be stronger. When added to fantasized construdo love creations, the beholder, with these drives in tandem, experiences the strongest love-sexual drive attraction that human beings can experience. Seeing the idealized beloved will involve construdo's disconnecting from certain reality aspects.

CHAPTER 2

UNIVERSAL FORCES OF PHYSICS AND BIOLOGY: THEIR EFFECTS ON EXPRESSION AND INTERACTIONS OF THE DRIVES

The drives not only influence most human behaviors, as do the drive derivatives, but also further influence multiple derivative interactions. The psychic drives derived from physics – construdo and destrudo – define boundaries of human behaviors; libido operates between them.

It is proposed that the organization of these drive-forces in earthly animate forms (e.g., in people) follows the precise organizational principles found in the universe. Thus, these forces or energies are organized in the macro universe will be the way they are organized within smaller systems in the micro universe (e.g., the systems of our brains).

THE UNIVERSAL ALTERNATING INFLUENCE

In the biochemistry of the cells in the human body and brain, the constant processes of building up and breaking down substances in the living molecules occurs: the anabolism and catabolism of cells. These processes coexist. Within the atoms composing the molecules in the cells of the body and brain, subatomic particles also alternate between building-up matter particles and anti-matter breaking-down particles, such as when an electron is created as a virtual particle pair with a positron. Yet subatomic particles within such a subatomic realm annihilate each other on occasion when they combine still further producing new particles.

These two sources (anabolism and catabolism) feed the fetal unconscious to register building-up influences and breaking-down influences. This is the biological substratum of the construdo and destrudo psychic drives' alternating as interplay in subatomic physics and therefore presumably present in all living things.

This opposition of forces in the two unconscious drives (construdo and destrudo) springs from the oppositional forces throughout the universe – the expanding forces; that is, building-up forces resulting from the Big Bang versus the gravity forces that would pull the universe back to a contracting universe – the breaking-down forces.

An alternating universal influence is observed in our solar system when sunspots alternate over an 11-year (or longer) period. Other alternating examples include heavy rain periods, which appear to alternate with the appearance of sunspots, and the growth of planetary flora is increased during these periods. Some scientists believe ice ages occur every 10,000 years; the earth's north and south magnetic poles reverse themselves, alternating every 700,000 years; and the earth's surface plates maintain equilibrium in their building up and breaking down, just as the cells in our bodies do. Again, building-up processes such as hydration synthesis and breaking-down processes such as dehydration synthesis occur within cell molecules and reflect another alternating influence. Universal alternating influences can be seen in the smallest organization of subatomic particles – in the atom – and of course, existing in all inert and living things on our planet.

In the ongoing attempt to arrive at a unified theory of the four uniting energies of physics, one interesting development is the appearance and formulation of string theory. A string is considered to be the smallest, non-divisible constituent of a sub-subatomic particle. It is theorized that strings show a constant oscillation of energy. The vibrating energy of strings – if proven – would be an additional influence of the universal alternating energy. However, the unifying universal force sought in string theory existed before the four differentiated forces from which TDT (Triadic Drive Theory) is derived ever crystallized into existence in our universe. This universal unifying force is prior to life and prior to the basic psychic drives that can be seen in all living things and inferred in the unconscious energies. Therefore, this unifying force is not pertinent as an influence on human beings or any other living things on our planet, because this unifying universal force existed prior to the life of basic psychic drives and of course prior to as well to the four differentiated forces of the subatomic world

The universal alternating influences are observed at all levels of organizations and systems. They affect the developing fetal brain by the time of six months in utero and, it is proposed, these universal alternating influences register in the fetal unconscious. The universal alternating influences are produced in the microscopic eighth level of system organizations in our universe as seen in the atom, in the system of organization of subatomic particles, as well as in still smaller particles of these subatomic particles and possibly in still smaller strings. This entire process of the presence of the universal alternating influence is consistent with the consideration that the psyche, which is but an infinitesimal speck surviving for a scintilla of a moment within our vast eternal universe, also presumably shows this influence. This influence causes drive alternations primarily between construdo and destrudo as well as their derivatives, and in addition, secondarily through their combinations with libido. The alternating unconscious influence underlies conscious drive reversals also observed by Sigmund Freud (1920/1955a) and by Anna Freud (1936).

Evidence of the alternating influence on construdo and destrudo, as well as their derivatives, can be observed in human behavior. For example, people create businesses (construdo). One's business may break down when its managers change outdated ways of doing things (destrudo) and insert new procedures (construdo). Workers may resist the new procedures (destrudo), then ultimately learn to adapt to the changes (construdo).

Let's look at some other examples. A political party's position is constructed and presented to voters (construdo); voters approve it. Then the position causes a reaction from the opposition that wants to tear it down (destrudo).

In history, groups of people have established civilizations (construdo); outside groups have observed them with envy and greed and a desire to topple them to be replaced with their own leadership (destrudo). If they succeed, in time another group will in all likelihood attempt the same process.

When peace is established a civilization may flourish (construdo). This condition, however, alternates with war and the breaking down of civilizations (destrudo) by those who find aspects of the civilization intolerable. The intervals of alternation for civilizations and war are determined by humans, so they are not the evidence of precise consistent intervals as are the universal forces of physics.

The Moods and Emotions

Drive alternations and drive derivative alternations cause psychic energy interaction or friction between these energies, as does their space-time direction. Drive aims are forward, while others are backward in space-time. The friction or interaction between the drive alternations and drive derivative alternations produces human moods such as elation, depression, happiness, anxiety, joy, and sadness. Moods persist longer than emotions, as emotions are transitory.

Human emotions are caused similarly by drive derivative interactions and by the friction between these energies: the different directions radiate out and away from a point in space-time in the case of destrudo derivative energies. These derivative energies discharge at different rates. The impact and disruption to the drives and consequent derivatives – by the event, in space-time – compounds the types of possible interactions, all contributing to producing emotions. Emotions result from less focused and less intense drive firings, as well as less prolonged drive firings and resultant interactions than is the case with moods.

In the derivative energy interactions producing the emotions over those drive energy interactions that produce moods, there is a refinement of energy or rarefied energy related to the emotions. Thus, as suggested, in the emotions the intensity and frequency of the drive is diminished; being happy, sad, fearful, proud, anxious, feeling inferior, bitter, envious, jealous, and so on, are more transitory than the deeper moods.

Anxiety can be both a general mood and a transitory emotion; it occurs when construdo conscious and unconscious anticipations, expectation, and future plans build up in the drive, but these are negative and destructive. People visualize themselves in a variety of anticipated actions and behaviors, but at the same time they visualize and expect these actions or behaviors to be broken down by destrudo impulses in themselves or others. People create both possibilities in their drive-differentiated unconscious (DDU), alternating between their connections with reality and their creative destrudo-feared possibilities – adverse to their positive fantasies and plans. The result is anxiety. Whether as a mood or a transitory emotion, anxiety is always produced by these construdo–destrudo fantasized alternations.

Triadic Drive Theory and Its Relationship to the Uncertainty Principle in Modern Physics

In 1927, Werner Heisenberg stated his "uncertainty principle": the more precisely the position of a particle is determined, the less precisely its momentum can be known, and vice versa. The uncertainty principle of modern physics negates the 19th century deterministic physics to which Freud wanted his psychoanalytic theory to adhere. Once Freud incorporated the science of his day into his theory, we find in his theory the doctrine of determinism: every thought or feeling (or Freudian slip) is determined by an actual cause from within the individual's unconscious.

However, TDT rejects the 19th century deterministic idea, replacing it with the 20th century uncertainty principle of modern physics. Thus, Freud's constancy principle and his principle of psychic determinism are not incorporated into TDT; rather, psychoanalysis in TDT is brought into congruence with ideas of modern physics. TDT bases its propositions on more recent scientific knowledge about the true reality of our surroun-dings and how this affects and determines what we do.

TDT is derived from the concept that humans exist in a four-dimensional world of three space dimensions and one time dimension. Hawking (1988) stated that to understand our existence, our world has to be considered as four-dimensional.

The Space-Time Dimension in Modern Physics

The space-time concept of physics is well suited for diagramming the past, present, and future for a psychological impulse of one of the drives: either construdo or destrudo. With space-time one can diagram how a current psychological event is derived from a past conscious or unconscious event and how these events will influence future events.

In TDT, space refers to a psychological event at a given place in one's life. Time refers to when the event occurred. The space-time diagram of any drive shows how the event occurred or where to look for trouble in the drive's course – that is, how its aims became fixated and the drive did not result in progressive effects. Clearly, the uncertainty principle as to "when" and "where," should be considered when making such predictions.

Incorporating construdo as a TDT drive also opens an entirely new concept in psychoanalytic inquiry: "How did an individual's past plans and fantasies for future space-time become disrupted by drive conflict, drive trauma, or drive arrest, yet continue to influence the individual's future directions as can be revealed in a space-time diagram?" The answer is that though construdo is disrupted in its forward flow, TDT proposes that such construdo remains with continuing unconscious influence on what the individual will do in the future.

By inserting construdo into the space-time diagram, contemplating where construdo has come from and where it is going, when the axes cross at the present event this axes crossing may have come from conscious, remembered precursors of that event or from unconscious precursors. It is likely that when these precursors are disrupted in the unconscious, or in consciousness drive, disruptions may occur in the present event.

TDT assumes that every individual has a normal energy transformation of the expended energy. The "replenishment principle" of psychic energy is akin to the "conservation of energy" principle of premodern physics, which states that energy cannot be created or destroyed.

However, Einstein postulated that energy could be transformed or converted into matter, which was the advent of modern physics. Similarly, Freud (1900/1953a) found that thoughts in dream transformations were like visual images in the unconscious. The TDT section on the unconscious will show postulate that operations of subatomic particles within the atom correspond to the operations in the unconscious mind. In TDT, transformations within the atom (in physics) are seen as tantamount to psychic energies of the unconscious.

The Exclusion Principle

The "exclusion principle" of modern physics states that two subatomic particles of identical spin, half spin particles, cannot (within the limits of Heisenberg's uncertainty principle) have the same position and velocity. According to TDT, the drives do not fuse, as long proposed by psychoanalytic theory. TDT asserts that drives fire sequentially. Although the time difference between these drive discharges may be infinitesimal, they are not concomitant. Conversely, TDT infers that drives do not fire simultaneously at the same target.

This means that object-striker destrudo, object-connecting construdo, and libido are not fused as suggested in the psychoanalytic concept of sadism. Rather, they sequentially express three distinctly different aims toward the same object or person. It also means that libido, self-striker destrudo aims, and self-construdo connecting aims do not fuse, as in masochism, but rather are sequentially different self-destrudo aims, autoerotic libido aims, and self-construdo aims, combined into one mood and emotion by the interaction of the three drive aims. TDT offers a radically different concept of the drive combinations and interactions than Freud's dual-drive theory. Without construdo, Freudian theory lacks a connecting force among drives that causes them to act and interact. At times, Freud uses the concept of cathexis as causing drive attachments to an object. Along with this catharsis, there is an increase in excitation in the neurons behind the drive's attachment.

Construdo holds the drives to their aims and objects. Without awareness of Construdo and destrudo as being in opposition, and with libido between them, Freudian theory could not indicate the relationship between the drives in terms of their positions relative to each other. While Freud's concept of narcissism indicates libido takes the ego as its object's aim, this concept does not consider a construdo drive force or influence uniting or connecting individuals with themselves. In contrast, narcissism in TDT is tantamount to being overly self-connected by construdo, beginning in uterine times with self-connecting uterine aims. This aim continues after birth. It predominates over the construdo aims to combine and connect with the environment.

PSYCHOANALYTIC PRINCIPLES SUPPORTED AS FORCES OF PHYSICS

The way the psyche has been observed to operate – how its drives' energies operate and interact – is attributed by TDT to the way the universal forces of physics affect the physical brain. However, this has not been shown to be a consistent effect explaining all the drive operations. In some operations of the drives TDT can find no known forces of physics to explain how the drives are operating. In these instances, TDT assumes that physics will find these forces. The following section presents these instances.

The Principle of Returning to Undermost Energy Origins

Stephen Hawking's (1988) theory that the universe will switch from Big Bang to Big Crunch posits a parallel between universal theorized forces of physics in the universe. Similarly, we here posit a theorized principle of TDT operating on the human psyche: that is "the principle of returning to the undermost energy origins." This principle indicates that exhibited mental energies in a present mental event ultimately return to that place in space-time from which the drive energy interactions originated. This is an example of where and how physics can theoretically support a triadic drive theory (TDT) concept.

Freud (1920/1955) observed in the drive energies a persistent tendency to repeat unsuccessful solutions to problems that the drives had encountered in the past but had not solved. When the conscious intent is to solve the reappearing problem, it is never solved, because the old solution is unconsciously reapplied. Freud called this the *repetition compulsion*, namely, a compulsion to repeat unconsciously the same failing solution to the earlier appearance of the same unsolved problem. This compulsion demonstrates the principle of returning to undermost energy origins.

Triadic Drive Theory accepts the accuracy of Freud's observations. The repetitive nature of highly emotionally charged situations is observable in human behavior. However, TDT attributes the cause of such repetitive behavior to mental forces derived from universal forces of physics rather than biological forces, as Freud assumed. Freud also saw it as related to his aggressive drive, but TDT does not. Nevertheless, there is a biological threat in the repetition, because the human mentality can carry the repetition. For example, an employee rebels against an employer in the same way he defied the authority of his parents in childhood. The employer notes the defiance and reacts negatively by firing the rebel. This is a repetition of the child's flawed solution during childhood. Originally, the parent punished the child's defiance, as the employer does now. Here is an example of disruption of a construdo aim to do the job in one's own way, coupled with destrudo intent to break the authority while still moving forward positively in the job. Yet the individual is compelled to repeat the destrudo solution that did not solve the original problem with parental authority, and once again causes the same result he experienced with his parents in childhood.

A similar influence can be observed in the creation of Earth's solar system. Four and a half billion years ago stars exploded (total smasher destrudo), leading to swirling gases and dust clouds that in time coalesced – combined and connected – building up the sun as a medium-sized star and forming the planets and their satellites into our solar system. The system has been constructed by gravitational forces that continue to hold the revolving planets around the sun. The solar system is a miniscule recapitulation of the evolving universe. Billions of galaxies were created after the Big Bang. Within the universe's unfolding energies is our solar system, a miniscule part that returns to the energy origins that created the entire universe.

Triadic Drive Theory assumes that parts of an energy system's fate will recapitulate the ultimate fate of the whole system. The Big Bang began or created our universe. An exploding star began our miniscule solar system, a part of the entire universe. Its fate of collapsing back into itself in the Big Crunch is a speculation of Hawking (1988) and also predicted by TDT. TDT believes this because construdo and destrudo conflicts are seen to return to their undermost energy origins psychologically.

There are other indications from the miniscule parts of the universe's energy actions that recapitulate what will ultimately occur in the entire universe's energy reactions, as seen as parts of the energy reactions in the galaxies of the entire universe. Galaxies contain black holes. Astrophysicists point out that these are stars that have collapsed back into themselves. It was further theorized that the gravitational field is so powerful within a black hole that nothing can escape it, not even light particles. (This conclusion has since been somewhat amended). Hence, they are seen only as darkness in observed galaxies. Black holes represent an infinitesimal part of the entire universe, but they operate by forces that ultimately will control the universe's operations, foretelling that someday the entire universe will collapse back into itself. In Hawking's Big Crunch, all the matter in the universe will be pulled back into itself by gravity, overcoming the expanding energy of the universe. This event will cause a great explosion, perhaps another Big Bang. What had followed might then follow again.

The effect can be observed in the psyche's tendency to repeatedly return to earlier psychic energy organizations. If psychic energies operate this way, the energy organization seems likely to be derived from universal forces of physics that continually repeat themselves. We see here an alternation between construdo influences derived from physics and destrudo influences derived from physics, forward-in-time influences versus backward-in-time influences. The alternation of these forces is repeatedly recapitulated. Such an event and the time of its happening constitute the space-time that is returned to.

These alternating universal influences – building up the universe to breaking down the universe – lend support to the oscillating theory of the universe. It indicates that it builds up, breaks down, then builds up again, and continues in alternations. In this theory, the Big Crunch becomes a Big Bang, creating another universe. While these examples from astrophysics can be seen as supporting the psychological principle of returning to undermost energy origins, physics has never advanced such a principle.

The Unknown Particle in Brain Biophysics

TDT conceptualizes that construdo energies combine and connect and moreover construct things from these combinations and connections. TDT also conceptualizes that construdo energies combine and connect with destrudo energies, as well as libido energies, by the process of construdo, to people, things, and events. This construdo energy effect also suggests from physics that there exists a particle interchange in the brain between the two energies construdo and destrudo. Physics tells us how such effects come into existence, but brain biophysics has yet to discover a particle that is interchanged in the brain. Until the appearance of the present TDT theory, biophysics has had no reason to search for such a particle; but if and when physics finds such a particle, it will add to the principle of TDT that predicts unconscious influences on the brain to reveal forces of physics underpinning it all.

Triadic Drive Theory and a Current Physics Controversy

TDT's alternating principle and the principle of returning to undermost energy origins are most applicable to the present controversy in physics regarding string theory (Green, 1999) versus the standard model of subatomic physics. Strings are described as vibrating, undulating energy that take the form of strings. Strings are thought to underlie all matter and interactions, thus explaining gravity and other forces, as well as particles – in everything, and everywhere. Therefore, the string theory has been called "the theory of everything." It is hypothesized that strings come in several conformations. Strings are so infinitesimal that they cannot be observed, leading to a controversy as to whether they exist at all.

TDT speculates that strings are an alternating energy state along with the matter state of particles. Einstein showed that the two states were interchangeable. TDT holds that as the universe operates, so will its infinitesimally smallest systems operate. Therefore, it is proposed here that our unconscious minds operate the way subatomic particles operate within the atom. Theoretically and presumably, the energy of these particles can be conceived as derived from the string energy underlying the subatomic particles.

THE UNCONSCIOUS

TDT theorizes that the principles of unconscious transformation of latent to manifest dream thoughts formulated by Freud are actually fundamentally caused by the operations of subatomic particles originating in the atoms of the fetal brain, and behave according to the parameters and limits of Heisenberg's uncertainty principle. It is further proposed that the influence of these operations causes the human unconscious to operate as it does. Thus the forces of physics at the subatomic level on the scale of the universe again, actually cause unconscious functioning in human beings. This speculation is consistent with the overall hypothesis of TDT that the universal forces of physics activate and cause basic mental functions.

Consider one of the parallels and theorized causes of unconscious operations from TDT: dreams are products of unconscious construdo operations. Freud discovered that thoughts underlying dreams (the latent dream content) were subject to condensation and fusion into a tight compacting of the thoughts, also underlying the visual images of the dream, those of manifest descriptive contents (the actual descriptive dream). Construdo's combining and connecting aims cause the condensation of dream thoughts into dense dream visual images. A telling analogy here is that modern physics shows that particles within an atom are compacted with increasing tightness by the property of confinement. "Condensation" as a salient feature of the psychoanalytic dream domain is therefore analogous to "confinement" as a salient feature of physics.

Freud also found that dream thoughts were subject to displacements of conscious thoughts and emotions. He indicated visual images representing these thoughts and emotions can be moved by dream processes from parts of the dream where they are connected to other visual dream images to places in the dream that seem utterly unconnected to the dream thought or image as far as placement and sequence of these thoughts or images appear.

In subatomic particle operations, a particle can be created as part of a pair, each located in two distinctly different places simultaneously. This seemingly illogical, irrational factor also underlies dream displacements. These so-called aberrations facilitate unconscious processes to conceal the dreams' true meaning from the individual's conscious comprehension. It makes the difference between the remembered manifest content of the dream and its disguised, deeper, latent unconscious content. This unconscious content of the dream's images is the real or actual meaning of the dream to its dreamer and concerns the unconscious nature of mental issues within the unconscious.

Freud believed the unconscious processes creating dreams (early differentiating unconscious construdo) to be timeless and spaceless. Subatomic particle operations from which it is proposed that dream operations are derived are timeless and spaceless, too, and operate from a different reality than what is logical in the conscious world. In contrast, it is assumed that reality is based on logic and reason. But Freud found that dream processes were derived from a different reality. This confusion of realities can be seen also with respect to physics insofar as a subatomic particle, for example, will change its identity from a matter particle to an anti-matter particle as it moves through what our conscious reason and logic would assume to be time.

In summary, Freud found that dream thoughts could be compacted, combined, and connected by the creative construdo aim in the symbolization of thoughts. That is another means of condensation, stemming from the confinement property in subatomic particle operations that begin in the fetal brain and continue in the human unconscious following birth. Thus, it may be proposed that the alternating influence of principles from physics on the fetal unconscious, as within subatomic particle influences, can alternate or reverse a dream thought into its opposite. It should be also noted that this reversing unconscious influence is far more extensive in dreams, and as a cause of psychopathology, than even Freud conceived.

Mental reversals into their opposites may be responsible for the processes that lead to psychopathology, because individuals who experience them never realize what is happening in their minds. Yet this is the way the unconscious mind operates. Individuals cannot understand why their minds are producing such reversed thoughts and feelings. Basically, people do not usually understand their psychopathology. Dreams deal often with past construdo fantasized influences, thoughts and emotions that a person has had. In the dream, the person wishes to redo these earlier construdo occurrences, recasting them in a more desirable way, a new, fantasized way, the way the individual wishes the fantasy actually had unfolded in the past.

Freud saw that a dream might be understood when it is considered to be a wish. He did not consider the construdo drives' influence in the past, and what the dreamer might have wished happened in the past. Freud did not consider that a wish determining or underlying a dream might refer to a past wish, and how the dreamer might have wished it had turned out. This way of thinking of such a process derives from the implications of TDT and is an example of how theory itself can generate new understandings.

A young child who is afraid may hate what it fears from the vantage point of being in a safe place. The child's destrudo impulses against loved yet feared that parents may cause unconscious self-destrudo created by guilt, built and reacted to by ideas of self-punishment. The child thinks nobody could love a child like itself, a child who at times hates its parents. Then the distorted reality perception may come, that other children also have negative feelings toward the child, a projection and displacement of such feelings from self to others' perceptions. Such internal processes (e.g., defense of projection and displacement) reverse reality. It is an important demonstration of how the reverse of reality can cause individuals to respond pathologically and, in this example, in a paranoid fashion.

Two Levels of the Unconscious

It is proposed here that our universe, solar system, planet, and subatomic particle operations, and possibly strings, influence the atoms of the human brain in the first six months of fetal existence, producing the fetal unconscious. TDT calls this the absolute unconscious (abs-unc). In space-time the influences of all the continuous forces converge on the fetal brain, resulting in intermittent impulses of drive energies breaking out from the abs-unc. This means construdo disconnects the drives and their different energies during the third trimester's fetal psyche that becomes the drive-differentiated unconscious (DDU).

This unconscious level is distinguished by the unconscious of the fetal brain before six months in utero; it is that of the abs-unc before three drives' energies (TDT) began to disconnect and separate from each other and differentiate as in the DDU. The uniting, binding energies (TDT) that construdo is derived from are markedly different from the universal energies that break things apart, like those energies underlying destrudo. Both differ from the life-recreating biological energies underlying libido.

Of course, the differentiating drives' energies (of construdo, destrudo, and libido) bear little similarity to each other. Construdo seldom combines and connects these energies. Construdo primarily disconnects them at this developmental time. Eventually these energies gather together by the grouping principles of proximity and similarity concepts as found for example in Gestalt psychology.

This grouping principle is the second developmental level of the unconscious, which it is proposed comes into existence in the human psyche during the third trimester. The grouping process also continues differrentiating and developing during the first six months after birth and differentiated construdo energies progressively seek to combine with the surrounding environment after birth. Just as all the drives' operations in the DDU are limited by the uncertainty principle, it can never be predicted how the drives will enter consciousness from the unconscious. Therefore, according to these implications, human behavior can never be completely and accurately predicted. This is quite different from what the Freudian concept of psychic determinism predicts.

Humans Seeking Reality

Because of construdo's inherent aim, construdo's combining aims continue in the new environment into which the newborn has been delivered. These combining aims also continue to combine with the self. In the first six months of life, unconscious perceptions from the abs-unc (absolute unconscious) cause the surroundings to keep changing visually. These perceptions show changes in shapes and colors; objects invert, and visual planes keep changing. Individuals begin to seek a stable, unchanging, perceptual environment, as the DDU begins to dominate their perceptions. Of course construdo alone will dominate in connecting them to the world.

The infant's construdo strives to accurately combine or achieve congruence between what it perceives and what is actually occurring in the surrounding new reality. Discrepancies between the two realities are discomforting and upsetting to the infant's comprehension of its surrounding new world. At this juncture the infant seeks sanity within its world, a balance with what is perceived and what is truly occurring in its new reality surroundings.

Planetary environmental gravity-reality also acts on the individual. It is the gravity attraction from the planet holding us to the planet. It sets for us our directions of up and down. The gravity force holds us to the realities occurring on the planet. In this sense the infant ultimately combines itself with the reality of our planet, because gravity's pull on humans is most constant and is most powerful in holding us to this reality. It should be noted that gravity is also one of the forces that created construdo.

Freud called the above listed construdo events the beginning of the reality principle. TDT considers this fact: the effect of construdo's striving to combine and connect with the stable and constant surroundings that are perceived to be there. That is different from the constantly cognitive changing reality Piaget (1954) observed infants dealing with, and called undifferentiated chaos.

Reality

As construdo perceptions from the DDU (Differentiated Drive Unconscious) begin to predominate during the first six months of life and get a hold on the environment, the abs-unc (absolute unconscious) perceptions recede back into what becomes the deeper level of the unconscious. Now the maturing infant is more settled with perceptions from the DDU, where each drive has its own hold on the environment in terms of its fundamental aim, which now realistically connects with the environment in terms of its differentiated energies (construdo, destrudo, and libido). The result is stable, unchanging perception and connection with the environment, far more preferable than the uncertain space-times of constantly changing perceptions that occurred during the first six months of life. Construdo aims lead individuals to combine and connect with what is actually in the world around them – our objective reality.

Psychopathology

In certain circumstances, drive energy regresses back to the DDU and the abs-unc. This drive energy may alternate between the DDU and the abs-unc, because this backward in space-time energy does not have the impetus in energy direction to progress into consciousness. Instead, the alternations produce psychosis; then we find psychotic perceptions of reality forming from these alternating unconscious levels as well as psychotic responses to the surrounding reality. The greater the frequency and amplitude or intensity of these alternations, the deeper and more entrenched the psychosis or mental pathology becomes, and the reverse is also true.

Similarly, in string theory, the differences in oscillations of a string in terms of frequency and amplitude result in markedly different consequences. In the subatomic and sub-subatomic particles, the frequency of a string oscillation determines which of the various elements the atom will be. The amplitude intensity, or space covered in the oscillation, determines the energy or charge of the particle in terms of a negative or positive charge. It would be premature to attribute to string theory operations in the differences in TDT with regard to types and depths of psychopathology, theorized alternations or oscillations between abs-unc and DDU, but string theory operations with respect to differences in TDT certainly must be kept in mind as possibly underlying the different psychopathologies.

Unconscious Construdo – Destrudo Creativity

Within the atoms of the brain, subatomic particles can smash into each other and thereby annihilate each other. These annihilations result in the creation of new subatomic particles proposed by TDT to also function in the abs-unc, which in turn influences the DDU, causing an individual to experience creativity as the energy moves back into consciousness under a new impetus to rejoin its construdo surroundings. This is the last stage of construdo. It also stems from the fact that within subatomic particles, sub-subatomic particles like quarks and antiquarks may combine and then annihilate each other, thus creating new sub-subatomic particles. Again, in Triadic Drive Theory influences of combining and annihilation lie behind creativity coming out of these destructive annihilations.

This effect can be observed psychically in the unconscious creation of dreams, in the creation of an idealized loved person, or in the psychotic self-creations of oneself as Jesus Christ, an approved ideal image, or the Virgin Mary. It might be the creative offering of an individual combining, connecting, and constructing from life's experiences a poem, play, or novel—a unique way to view reality as seen by the artist. It might be the way musical notes are combined and connected in the construction of a sonata, symphony, opera, or rock song.

The DDU (Drive Differentiated Unconscious) assembles discrepant parts of reality in a puzzle of how these parts might fit together uniquely. The conscious mind seeks the solution to the puzzle. The answer reemerges into consciousness as a "eureka" experience, and the creator wonders where the solution came from. The creator might be a theoretical physicist, mathematician, or economist. In all fields, the creative process is similar.

A patient who was a successful Hollywood writer, once commented, "Suddenly I experience creative explosions," then ideas flowed from his pen. According to TDT, he was putting into words the process of particle annihilation, destrudo; particles smashing into each other, resulting in new particles, which were combined and connected anew in the construction in the DDU (of a new idea or creation that reemerged back into consciousness). The individual often cannot understand how the result was produced in the unconscious leading to the creation in consciousness. Consciously, the individual experiences a wondrous creation from within the self, which comes into the individual's mind as a gift from the cosmos.

That is the influence the abs-unc has on the DDU: destrudo smashes out of existence the wrong solutions, and then allows the correct creative construdo solution to come back into consciousness. The result is a

31

completely new conscious idea, an original, unique way to conceive of the data, which the creator had never thought of previously.

THE DRIVE IDEAL

From four and a half to six and a half years of age, construdo and destrudo develop to their final stage aims, just as libido does. By age five, the three drives have at least reached their ultimate stage aims in development, although these may not necessarily have been completed. By the fifth to sixth year, construdo has combined and connected these divergent drives in constructed-drive conflict. Over time, construdo has conflicted with destrudo, then libido with destrudo, and then construdo with libido.

When all the drives have developed normally to their ultimate end stages (completed by age six and a half), many children face an over-whelmming triadic drive conflict for the first time. Most children's drives continue to develop forward in space-time without any disrupting drive conflict that would result in drive imbalances that throw drives out of equilibrium. Drive disruptions and imbalances occur when a child seeks an exclusive libidinous relationship with the parent of the opposite sex. But the child's construdo has connected and constructed a loving, binding relationship with both parents. The child's destrudo aims can be to topple, strike down, or smash out of existence the same-sex parent – the classical Oedipal conflict. The triangular drive conflict can be overwhelming for a child. At the very least what's inside itself, the contradictory inclinations, inner private thoughts, and DDU fantasies about triangular relationships are puzzling or even frightening to the child.

The strange directions in which its drives have taken it shock the child. It may sense that neither parent accepts its drive inclinations with respect to the Oedipal triangle. The child becomes consciously aware that society, religion, teachers, relatives, and neighbors do not accept its drive inclinations. These standards of others, contrary to the child's drives, might prevail, and sometimes do. But these drive's urges and motives are the essential part of the child, placing it in conflict with its condemning, rebuking surroundings. When such a conflict happens, the child becomes cautious about the situations into which its drives have led it. The child's developed construdo can lead it in several ways: to follow the directions of its parents and surrounding adults; to follow its own inclinations; or to put together influences from both sources.

These circumstances are caused by discrepant influences from its biological drive, libido, of which it is only vaguely aware. Further, there is destrudo to consider, that negative force from physics that affects everyone, which we only vaguely perceive in ourselves, believing it to be a human biological influence (which it is not). Destrudo is an influence from physics to break things down or break things apart so that they can never exist again. Finally, there is construdo, the uniting influences from physics of which we have no understanding or conceptualization, only sensing its influence as constructive humanitarian (construdo) urges.

These drive's influences on humans that are from physics rather than biology give us real reason to deal with the unclear, ill-understood, influences to which we are subjected. Such a discrepant set of circumstances creates the need for a psychic construct that can balance drive impulses with environmental requirements. The difference between a determining drive of biology and the more behavioral determining drives from physics causes construdo to seek a balancing mechanism between these two little understood, divergent influences.

A part of construdo, sensing the discrepancy, disconnects from the rest of construdo, because of the influences from physics directly on an individual emanating from all the universal levels, which are quite different from the biological influences. Some mediating psychic construct is needed to give closure and reconciliation to the conflicting drive influences from physics and biology, which are paradoxical. The *drive ideal* is the resultant constructed construct. TDT borrows from Freud's ideal concept (this ego ideal) that Freud called the superego, because it implies that people have ideas about what they would like their drive combination to be. It obviously stems from construdo.

Both construdo fantasy and conscious construdo thinking and judgment are made about how the drive impulses should cause a child to behave. With the emergence of judgment based on past results through time, choices occur among the drive aims. But which aims will lead to the greatest satisfaction between self-satisfying aims and environmental acceptance and approval?

The drive ideal causes us to construct families, communities, societies, and civilizations as we are driven by construdo to combine and connect with each other. When the drive ideal is shared with others it convinces us that we are on the right track, so we construct organizations with others. As each level of organization is achieved, construdo impels a person to join larger and larger groups. The drive ideal also tells individuals when their object destrudo aims should be applied to others who oppose the progress of such organizations.

Once these higher level organizations are established, some people will want to oppose them, wishing to destrudo defy the newly established order. What is built up by construdo influences may also be broken down by destrudo influences. Their different drive ideals tell them that their position is the right course. New groups come into existence; they can be united by shared drive ideals – the new group displays object destrudo, or shared aim to break apart, damage, or topple the established organization's position.

There is no observable alternative in construdo's building-up or its constructing new organizations, a positive way people connect with each other, versus a breaking-down, destrudo alternation, opposing these constructions. The interplay of construdo's and destrudo's struggle for dominance reveals their universal interplay. These alternations are determined by abs-unc influences on the DDU (Differentiated Drive Unconscious).

The psychoanalytic idea that some destrudo energy must be expressed to achieve good mental health is modified by the TDT concept that a conscious drive ideal choice of a substitution of construdo expression for destrudo expression can lead to mental health satisfaction, as well as energy replenishment to the DDU. Drive energy from the abs-unc and the DDU are interchangeable, allowing substitutions of one drive energy for the other.

It's a different issue when destrudo arousals and reactions have been held back, suppressed, or repressed. This destrudo energy must be released and expressed to acquire equilibrium for good mental health. Beyond newly expressed destrudo, when released these liberated mental energies employed for this purpose (for newly expressed destrudo) make energy available from the DDU to react with the environment. TDT's position is that substituting construdo energy for destrudo leads to a mentally healthy release of built-up psychic energy. This, the entire theory of triadic drive with respect to its influence in the psyche is a new understanding of the influence of physics on human behavior.

The unconscious component of the drive ideal is created in the DDU. It mediates what conscious results of the drives' aims will be approved by this organization of three supra-drives, determining what our drives will cause us to do, as well as what we will punish ourselves for doing or not doing. Self-punishment predominantly carried out by self-striker destrudo from the DDU can cause a striker attack on the self (perhaps even a sudden heart attack). Self-punishment from the DDU can cause perceptual connections with the environment to be unconsciously suspended, so that an icy path ahead is not seen, which facilitates slipping and falling, breaking an arm or leg, meeting the self-striker destrudo aim for self-punishment. Self-destrudo toppler aims in the DDU (Differentiated Drive Unconscious) leading to self-punishment can be seen in self-toppling, such as perhaps even in the weakening of the self's immune system, then allowing bacteria, viruses, protozoa, or parasites to cause infectious illness.

CHAPTER 3

THE ORIGINS OF THE CONSTRUCTIVE DRIVE

THE COMBINING, CONNECTING, CONSTRUCTING, CREATIVE DRIVE: CONSTRUDO

The progression in nature from one-celled living organisms to multicellular organisms is noted by Freud in *Beyond the Pleasure Principle* (1920/1955a), *The Ego and the Id* (1923/1961c), and *Civilization and Its Discontents* (1930/1961a). Freud was observing the progression of living cells to combine with other living cells into more and more complex organisms. Human beings are the final result or ultimate cellular organization of this progression. The progression reflects Darwin's (1859/1972) theory of evolution; the progression from lower to more complex organization of living organisms, which must have impressed Freud, whether or not he could say what that progression reflected in terms of the human psyche.

Darwin's biological theory of evolution and Freud's biological theory of human psychic drives show a wondrous progression toward higher, more complex organization of life. Freud and Darwin did not know what caused an individual cell that had achieved the quality of life to hold itself together, nor could they explain what caused living cells to bond with other living cells in a way that created progressively more complex organ systems, or why these organ systems held together as one body, resulting in progressively more complex organisms. Darwin appears to have taken it as an aspect of biology of living things. Freud appears to have taken it all as an inherent part of his life drive, libido.

Triadic Drive Theory (TDT) distinguishes three aspects of the process causing the progression to more and more complex organisms. First, there are the forces of physics billions of years ago causing things to combine, resulting in atoms fortuitously combining to produce molecules and then living cells. Once living cells were created, biology began. Second, there came a biological force to recreate living cells, to recreate themselves. These one-celled organisms reproduced themselves. Third, the forces of physics were operative in causing the recreated living cells to combine with one another, gradually producing more complex organizations of living cells. Over millions of years, after the first living cells were created, cells differentiated themselves. Again, this was a biological process, perhaps due to a series of genes causing adaptation, or survival of the fittest, or other factors. These differentiated cells combined with each other, under the force of physics that combined and held things together, producing differentiated groups of cells or different tissues. These differentiated tissues making up different organs in the living organism were the result of what Darwin (1859/1972) called natural selection. We see here the force of physics and biology operating simultaneously, in the progression to more complex living organisms, ultimately leading to the creation of Homo sapiens.

Without biological forces, life would not have recreated itself. Without the forces of physics, living cells and tissues would not have been held together or combined. Life could not continue without both. Human beings are essentially created by the forces of physics in the universe. Biology is an accidental presence that gave them life. Human beings there-fore sare an accidental miracle produced by universal forces.

The sun of our solar system is a medium-sized star in the galaxy that contains some hundred billion stars, and our galaxy is only one of some hundred billion galaxies, each also containing some hundred billion stars (Stott & Twist, 1995). Indeed, we are fortunate to have achieved existence in the incomprehensible vastness of our universe.

The drives of the human psyche that arise from these forces are different from those arrived at by Freud in his dual instinct theory. In *Three Essays on the Theory of Sexuality*, Freud (1905/1953b) advanced libido theory. The libido drive was derived from biology. Its ultimate goal was to reproduce human life. His concerns with human death in *Beyond the Pleasure Principle* (Freud, 1920/1955a) conceived of a destructive drive as the drive that opposed such combining, connecting forces of nature or physics. So too, Anna Freud (1972) wrote that aggression has the aim of undoing such combinations and connections in nature. Destrudo is conceived by both as a drive that breaks apart the combining and connecting forces of nature. Its ultimate aim is the total destruction of an object by breaking it apart into pieces.

The combining connecting forces of nature were considered by Sigmund Freud (1920/1955a; 1923/1961c; & 1930/1961a) and Anna Freud (1972) to be part of the influence of libido, the biological drive of life. Hence the living organisms on our planet reflecting the evolutionary influence of biology alone are seen as the outgrowth of the libido drive. Thus both Freud and Anna Freud thought that the unification, preservation, and propagation of life (in humans and other organisms) was an aim and an accomplishment of the libido drive.

TDT finds that both of these psychoanalytic theorists overempha-sized the role of libido in seeing it as the influence that unified and preserved life. Libido is the drive that propagates life. It does not preserve or continue life. Libido is the biological drive in the human psyche directing human beings to come together sensually and sexually, for the hidden, not consciously considered purpose of recreating the human species. The drive's aim is obscured from individuals participating in the behavior leading to the drive's fulfillment as had been observed by Freud (1915/1957b).

But there is more to life than merely its recreation. Human growth, the adding and combining of one living cell to another, the connecting of different types of cells with each other into human tissues and organs, and all of it holding together as a human body, involves the influence of forces different from the forces that recreated life. The former (connectedness) TDT attributes to forces of physics, the latter to biological forces (recreation of life).

INTRODUCING A THIRD DRIVE

The uniting forces of physics on the human organism give rise to a third drive in the human psyche, different from libido and destrudo that aims to combine things, or hold them together, to connect things that seem to go together psychologically (e.g., by their similarity or proximity to each other). This is the psychic energy that constructs from a variety of perceived different objects how these objects might go together to construct an object different from any of the individual objects themselves. It is the psychic drive behind putting together stimulation of the senses from reality impressions, or imaginable impressions of what a reality could be, into combinations or compositions that are unique or original, that delight and please our senses by the way the combinations or compositions are put together and the way they stimulate us.

It is our response to the genius of composition in art, music, poetry, prose, sculpture, or architecture. It is the wonder we feel by the way our sense are stimulated when the subject being presented to us is unpleasant and repulsive to us, yet fascinating. In the composition of the painting we would have never imagined we would be so gripped by the presentation of such unpleasant material. This being gripped by such artful presentation is the creative height of combining things. It's the creative stage of the drive.

Creativity can show itself to our visual senses in the way a painter puts together aspects of our visual reality in a way we had never seen before. We can hear such originality in a way a composer puts together notes and sounds, when we could never have imagined that these sounds would go together, to make the impression they have on us. Creativity can be found in the way a master chef puts together certain tastes, smells, and their visual presentations to produce a response of gastronomic delight.

Creativity can be found kinesthetically in the way a ballerina puts together various physical movements of her body, which we see and identify within our own bodies kinesthetically. The total combined effect on our senses produces in us feelings the ballerina conveys, such that our visual, kinesthetic, and emotional combined response leads to bodily emotional feelings that we have not felt before, and the originality of our response grips parts of us. A jazz or hip-hop dancer can produce similar reactions in us. We know we are being touched by the best of this drive in us – namely, the creative level of the drive, through the efforts of another person, and the composition of their work.

THE CONSTRUDO DRIVE'S CREATIVE LEVEL IN OBJECT RELATIONS

Falling in love is one way we combine ourselves with others. We see, hear, smell, experience touching them, or their touching us, perceive the way they move, or their achievements in life, or their particular knowledge is combined and connected to construct in our psyches a totally unique and original impression on us. The drive has put these elements of our impressions together in a creative way, not always immediately perceived by others around us. The creative impression or idealization of the object we have come to experience as love of the object. Then love is one of the creative levels of this drive in us.

TDT calls this drive of the human psyche *construdo*. Construdo is combining sense impressions, ideas, affects, kinesthetic reactions, and attitudes, to connect them in some pattern, to construct them into a definite entity, to create from them a unique and creative conceptualization. The term construdo gives emphasis to the fact that the constructing level of the drive is from the median range of the component aims of the drive sketched out thus far. It is a drive in us whose aims are life augmenting, constructive and positive in regards to our existence in the world, in opposition to the destructive aims of destrudo.

THE ORIGINS OF CONSTRUDO

It is time now to state the binding, uniting forces of physics influencing all things in the universe, the bodies of human beings, including their psyches, from which construdo originates. This is a departure from Freud's concept that the libido drive is responsible for the continuance or perpetuation of human life. The construdo drive is more involved in the perpetuation of life than the libido drive. Construdo is the primary drive that continues human life.

38

The four forces of physics crucial to the exposition are (a) the force of gravity, (b) the electromagnetic force, (c) the weak nuclear force, and (d) the strong nuclear force.

Gravity holds our bodies, alive or dead, to the earth. It is the force holding our solar system, our galaxy, our entire universe together.

The *electromagnetic force* holds together living molecules, cells, tissue, connective tissue, organs, organ systems, and our skin, binding our entire bodies together. It holds us together, as well as everything else on our planet.

The *weak nuclear force* holds together the nucleus.

The *strong nuclear force* holds together the subatomic protons and neutrons in the atomic nuclei in our bodies, as well as the other nuclei of atoms on our planet, and our universe (Davies, 1982).

Such information from physics was not part of the general knowledge in Freud's time. Such information was not likely, therefore, to have influenced his thinking about the drives. Nevertheless of course, changing times produce changing knowledge and theories.

The existence of these four forces, the effects they have on binding us to the earth, holding the minutest parts of our bodies and brains together, permits the assertion that these influences on our bodies and brains produces an unconscious experiencing of these influences from every cell throughout our bodies and brains. This influence, while never reaching the level of conscious thinking, has theoretically nevertheless had a formative influence on our unconscious psyches and a determining influence on how the unconscious creeps through into the conscious psyche to determine conscious behavior.

TDT perceives unconscious influences in creating and determining the drive behavior, with all its derivatives, in a third drive in human beings, to combine with psychic events, people, and representations to connect events and representations, to construct things from these representations, as friendships, associations, and relationships, or put them together, or combine them in ways unique to the individual combining them – that is creatively. These influences from the forces of physics have created the third drive in the human psyche – construdo.

DEVELOPMENTAL STAGES OF CONSTRUDO

The developmental stages of construdo can be observed in the different kinds of play children engage in as they grow. At first they gather their toys together, play with a group of toys, or put their toys together on a couch, behind a couch, or on their beds. This is the first stage of construdo development, the ***combining*** stage. It occurs between six months and one and a half years.

Next, they begin to connect their toys in some way. A doll is put in a wagon. Blocks are piled on top of one another. A ball is put in a cup. Such play continues from about one and a half to two and a half years of age. It is the *connecting* stage of the construdo drive.

From two and a half to four, children begin to construct things with their toys for the first time. Blocks are put together to build a bridge. A doll is dressed with clothes. An erector set is used to construct objects. It is the *constructing* stage of the drive.

By four years up to six and a half years, children's play becomes more imaginative and creative. Now they draw or paint pictures of the world as they see it or imagine it could be. They play out scenes of dreams with imaginary companions and with their friends. They imagine a green sky and an orange earth. They play at being an astronaut, a nurse, a doctor, a mother, a father, a teacher. It reflects the *creative* stage of the construdo drive in its development in these young human beings.

Evidence of the Construdo Drive in the Animal Kingdom

According to Darwin's (1859/1972) theory of evolution, in the animal kingdom, there is evidence of the construdo drive among the living organisms from which we evolved. The effects of the forces of physics undergirding construdo are seen in all other living creatures created by these forces. Insects, mosquitoes, flies, bees, hornets, swarm or fly together. Ants and cockroaches stay together. All varieties of fish swim together with their own species. Amphibians such as frogs or lizards group together. Bats flock together. All varieties of birds fly together. Mammals, from the smallest (mice, weasels) to the medium sized (zebras, antelopes, lions, gorillas), to the largest (elephants and whales) – all show an observable tendency to remain together with their own species or to ***combine*** with them – the first fundamental stage of construdo.

Different species all show this tendency. From the smallest brain or no brain at all, as in the case of the paramecia, microscopic organisms, to the largest mammal, the whale. The grouping together of the different species shows the influence of the drive to combine; they do it with what is living nearest to them and is most similar to them.

Similarly, when domesticated pets, such as fish, birds, cats, or dogs, show a distinct recognition of our presence and appear to want to combine with us, they are showing the influence on their brains from the forces of physics to combine with what is living around them. We become the object with which they sense or experience the drive to combine in their surroundings, particularly that which is living as they are. A pet dog, for instance, may follow us from room to room. But is the dog's behavior anything more than seeking to stay in proximity and contact with us or to be combined with us? We may anthropomorphize our dog's behavior as love for us, but is that really what our dog is doing?

TDT posits that the dog is displaying a persistent drive to remain combined with a figure that has become a living, moving part of its immediate environment. The behavior of other species indicates the same drive influence from the forces of physics, in the universe, and operating on all living things on our planet to combine with things around them.

Construdo in Plant Life

Can it be denied that a given type of plant life will show attempts to grow in all directions around and implanting of this life? Will not weeds overgrow or topple flowers? Will not grass overgrow other plants, such as dandelions? This occuring confirms the energy in other life – plant life. Construdo tenets cause us to expect that the laws governing other life will also apply to plant forms. These plant forms will combine, connect, construct, and create new plant forms. The construdo drive operates equally on all life forms.

More on Construdo in Humans

Construdo also causes people to combine themselves together and connect with each other in a variety of interpersonal relationships. The inherent nature of the drive to bind and connect is responsible for the connections of mental representations, mental events and occurrences, imagination, sensory perceptions, imagery, elements of ideas and concepts, memories, feelings, aspirations, fantasies, and so on in the human psyche itself. It is the energy of the drive that pulls these mental transactions together as mental organizations.

Construdo in Other Psychologies

Gestalt psychology addressed the issues of why certain visual stimuli were pulled together to form visual organizations, the principles of grouping visual stimuli into perceptual organizations (Koffka, 1935), as well as how these principles applied to social interactions (Asch, 1952). Classical conditioning experiments considered what happened when a stimulus that naturally and automatically produced a given response was paired in presentation with a second stimulus that ordinarily did not produce that given response. After many repetitions of the paired presentations, the second stimulus alone would produce the given conditioned response (Mowrer, 1960).

In psychoanalysis, the concept of cathexis indicates that affect, and psychic energy is invested or connected to an object, emotion or idea. In "The Neuro-Psychoses of Defense," Freud (1894/1962) referred to a powerful idea being weakened by diminishing the affect invested in the idea. He also writes of the total excitation that is loaded into the idea.

In the preceding paragraphs, visual groupings of stimuli, conditioning of stimuli, and attachment of psychic energy and emotion to stimuli are found to establish mental unities in the human psyche. Where does the psychic energy come from that holds these psychic entities together? Why does the energy operate in such a manner that it causes mental events to unite or connect with each other? We have simply tended to take for granted that our psychic energies operate in an additive, combining manner. But if they did not, then we would not be able to add memory to memory, emotion to memory, concept to concept, memory to concept, or emotion to concept. We would not be able to learn or develop mentally. There would be no complexity to our thoughts, emotions, or perceptions. Nothing would come together in our psyches. Mental events would simply remain apart and disconnected. But they don't. Construdo energies join or add together mental phenomena or happenings in the psyche.

The physical energy of the uniting forces of physics (gravity, electromagnetism, the strong and the weak nuclear force) in the cells of our bodies, and in the atoms of our brains create a mental energy in us causing us to react similarly psychologically to things around us, that is, to mentally unite them.

Construdo's Disconnection from a Bond

Now to the matter of the construdo drive reversing or disconnecting itself. Both Sigmund Freud (1915/1957b) and Anna Freud (1946) observed that the drives often reverse themselves in the unconscious. We have already indicated how the energies of subatomic particles show reversals of particle identity as a parallel of energies that determine reversals of the drives in the unconscious psyche.

We often see in psychoanalysis a person professing a sexual revulsion for another, while in the unconscious of the same individual there is an intense sexual attraction to that person. Similarly, we may hear the strongest positive, loving feelings one has toward another, and yet find in the unconscious of the same individual repressed hostile feelings toward that individual.

Freud (1911/1958b) observed that libido could be withdrawn from people surrounding a person. Hartmann (1964) wrote of the neutralization of aggression. So both Freud and Hartmann were indicating that the drives may be withdrawn from objects and ideas.

In terms of conditioning phenomena, behavior modification therapy demonstrates that a stimulus can be desensitized, so that it does not produce an undesired response (Wolpe, 1974). The conditioning process is reversed. Thus, construdo, the drive to connect and unite mental happenings, can reverse itself and disconnect or uncouple mental events in the human psyche. It is the disconnecting energies of construdo that break the connections.

People can feel connected in parent-child relationships, family relationships, love relationships, business relationships, community organizations, or friendships, and then suddenly disconnect because of newly evoked destrudo. They then feel disconnected and detached from these people with whom they had formerly felt intensely connected. Destrudo reactions frequently break apart the construdo bonds.

Newly formed construdo connections can replace old connections because the objects are more appealing or the ideas are more convincing and compelling in a new connection that disconnects from an old connection. Newly formed libido connections can replace old ones and create new lustful bonds. Construdo can be reversed, therefore, by its own unconscious energy reversing itself or with energy from either of the other two drives (destrudo and libido). Drive fusions are the result of the energy of construdo being used to unite other drives. The fusion of libido and destrudo involves the energy of construdo uniting them.

CHAPTER 4

DEVELOPMENTAL STAGES OF CONSTRUDO

CONNECTING AND DISCONNECTING WITH SELF OR
OBJECTS: CONSTRUDO STAGES AND THEIR DERIVATIVES

Uterine Times

The influences of the combining, binding forces of physics on everything in the universe are external and omnipresent. Therefore, the axiom becomes that their influence on the developing embryo is immediate. The first stage of influence from these binding forces is to combine what has come into existence in the universe. Thus, in the reproductive process, as the egg cell from the female parent is fertilized by the sperm cell from the male parent, the tendency to combine is evident.

THE FIRST INTRAUTERINE STAGE: SELF-COMBINING
(ONE TO SIX MONTHS IN UTERO)

The combining of the egg and sperm shows the influence of the universal binding forces on human beings. The fertilized egg cell divides into two cells (a biological process), and these two cells hold together (a process of physics); each of these cells divides again into four, then eight, and then sixteen, and continues dividing in this fashion. The human embryo thus comes into existence. At three months of uterine development the fetus has attained its basic structural plan, as differentiating cells form different organ tissues (a biological process) and binding together (a process of physics); at six months, the fetus has attained the essential human form, organs, and proportional body shapes and relationships.

The sensation of body skin folds against body skin folds creates a sensing in the fetus of touch self-contact, self-combining physiologically and sensually through contact with its own body. Moreover, there is the touch contact of the fetus's body with the walls of the mother's uterus. Touch is the first sense human beings have of being combined with themselves, as well as something else – the mother's body. Sensing touch with oneself facilitates the sensing that one can combine with oneself. It is registered in the human unconscious. This is the self-combining stage of the construdo drive. It is the beginning of the human progression toward independent life. By six months of uterine development, there is a nebulous, unformed, rudimentary sense of self-identity, different from the uterus and its fluid environment.

During these first six months of uterine development, the embryo and fetus may sense unwelcoming or rejecting feelings from the pregnant mother. The mother may wish she were not pregnant; the baby may cause the end of a desired career or the loss of income. It is possible according to Triadic Drive Theory (TDT) that the fetus may sense these ambivalent feelings. Thus, instead of progressing forward to the sensing of being connected with the mother's uterus, the fetus may recoil correspondingly back into itself, and experience regressed sensings of being combined with itself, as in the folds of its skin against its other folds. This sensing of being combined with itself may continue throughout the rest of the pregnancy, and it may continue after birth. In such a case the individual becomes overly self-combined and self-connected.

Due to unfavorable conditions in the uterus a similar situation can occur when the pregnancy is in danger of spontaneous abortion. The embryo/fetus may regress to being self-combined. Even if it survives the threatened miscarriage, the embryo may still have strong sensings of its self-combining and self-connectedness.

One man who survived the mixed feelings of his mother toward her pregnancy reported that she later in life told him she wanted to commit suicide while she was pregnant. Nevertheless, this man showed a great reliance on himself and his perception of reality. He did not show characteristics of being overly self-connected. Rather, he was attracted to women who displayed a characteristic with which he was intimately involved in utero and in growing up: namely, women who considered their own interests before others. He found in them characteristics he consciously denied in himself, but yet to which he was either apparently or presumably unconsciously attracted.

THE SELF-CONNECTING AND SELF CONSTRUCTING STAGE (SIX TO NINE MONTHS IN UTERO)

By the third trimester of pregnancy, all of the organ systems of the fetus's body have been formed. The growing fetus adds mass to its developing body. It develops a sense of being connected with the uterus and its fluid environment. The sense of connectedness is, according to TDT, the result of the impact of the universal, eternal, binding forces of physics operating on the fetus's brain and sensorium. The fetus senses its body in being self-connected. In this sense, the growth processes add to its psychic sense of its self-connectedness by unconscious registering of its body occurrences.

46

Implications regarding TDT and its undergirded basis in physics suggest that the fetus's developing brain can experience hallucinatory sensations. Reasons include the strong and weak nuclear forces on the atoms of the brain, along with the electromagnetic forces that hold the brain together along with the rest of its body. Gravity, which holds the fetus in its uterine position, also plays a role in stimulating the brain during development. The brain of the fetus is being stimulated (or fired) by the universal subatomic operating forces; the fetus has no control whatsoever over these stimulations or how they affect its elementary brain.

Toward the end of the third trimester, the fetus begins to sense the construction of itself, as a thing or entity. It has some vague sense that it is distinct and different from its uterine environment. When it moves, kicks, or moves its hands and arms, its uterine surroundings do not move. Rather they remain stable, offering contactual resistance to its actions.

These interactions cause the fetus's sensing that possibly there is a difference between itself and something that might be surrounding it. At this time it certainly does not consciously know this or have an understanding of the difference between itself and its surroundings. It has been constructed by biological forces and held together at each step along the way by forces of physics; most importantly, it is developing a psyche or mentality from these forces of physics. Its construdo drive is causing it to mentally connect with its environment the way it is connecting with itself, although certainly, as suggested, there is no awareness of this drive influence. The fetus's psyche is operating as the unconscious operates. Yet in fleeting, momentary instances the fetus senses a difference. Somehow it is a constructed entity, separate from the uterine surroundings. This minute difference then prompts construdo to more precisely combine and connect with what is around it – the uterus.

SELF-CREATING STAGE (BIRTH)

Birth causes the infant to experience radically new sensations and perceptions. The infant erroneously senses that it hascreated all that is now new. The just-born unconscious psyche feels the experience of life outside the uterus as its own creation. It made this new place. There are vestiges of construdo and self-constructing aims here. What construdo has added are self-creation pathways, erroneous ways of sensing the new reality surroundings, or false connections. The infant senses that it has created a new reality, but it has not actually created anything. The just-born is only responding to the world outside the uterus into which it has been delivered. What happens is that the infant's unconscious registers that it has created itself.

The Unconscious Reflections of These Forces

Time, space, and logic are completely outside the realm of these stimulations on the infant, or their registering on the rudimentary, unconscious psyche. The fetus's rudimentary brain and psyche have no control over these stimulation firings along with no ability to blot them out.

Freud's (1900/1953a, 1915/1957b) discoveries about the operations of the unconscious parallel what physicists later discovered is the nature of the operations of the subatomic particles in the atom. Freud did not explain why the unconscious operates as it does. He did, however, describe its operations.

In Triadic Drive Theory (TDT) the unconscious operates as it does because the principles that govern the operation of subatomic particles in the atoms of our brains create a field of influence in the fetus's developing brain.This occurs during the second and third trimesters of uterine growth, causing the fetus's rudimentary psyche to begin operating in a way similar to the operation of subatomic particles. There is little else influencing the psyche at this time. Thus we can even poetically envision that in our psychic development of the unconscious, we become creatures of the universe. TDT extrapolations consider that our first mentality is truly created by these universal forces.

The first level of organization of the rudimentary psyche remains unconscious, as the fetus has not yet entered our reality. Of course, after birth, the infant has entered a world of stimulation and shows responsiveness. It is a responsiveness that draws the infant's psyche into the world of reality based upon the psyche's construdo's influence.

Freud's discovery of the operations of the unconscious mind included displacement in the unconscious of a thought or affect from one place to another place, often at the same time; substitution of one idea or affect for another, in symbolization, or the reversal of a thought or emotion; and several ideas and affects being compressed into one idea or affect, as in a condensation. Freud found the unconscious was timeless in its operations, just as the operations of subatomic particles are in the atom. Spatial relations have no meaning in the unconscious or in subatomic particles. Moreover, in the unconscious there is no logic as we know it, no logical contradictions or negations. Similarly, subatomic particles operate without logic. The uncertainty principle in subatomic physics indicates an unsolvable problem: when a subatomic particle's (always moving) position within the atom can be determined, its velocity or speed cannot be determined, and vice versa. This is not what our conscious logic or reason would expect! This discovery is a prelude to the discoveries of physics about subatomic particle operations in the atom. Freud's division of the psyche into conscious and unconscious parallels the fact that in the atom, every subatomic particle has its antiparticle (Chester, 1978). Similarly, every conscious reality, thought, and emotion has its unconscious counterpart.

For the fetus at six to nine months in utero, logic has no meaning in the imagery flashing through its brain's time and space. Freud (1915/1957b) was explicit on these aspects of the unconscious. He asserted that the processes of the unconscious are:

> "...timeless, i.e., they are not ordered temporarily, are not altered by the passage of time; they have no reference to time at all. Reference to time is bound up, once again, with the works of the system "Cs" (conscious). . ..When a process passes from one idea to another, the first idea retains a part of its cathexis and only a small portion undergoes displacement" (Freud, 1915/1957b, pp. 187-188).

Thus, spatial positions have no meaning or exactitude in the unconscious psyche. Moreover, in the unconscious there is "...no negation, no doubt, no degrees of certainty... Negation is a substitute, at a higher level, for repression" (Freud, 1915/1957b, p. 186). In other words, there are no contradictions based on rational logic in the unconscious.

CONSTRUDO IS BEHIND THE ENERGY OF CATHEXIS

In Freud's (1915/1957b) account of the unconscious, he frequently employs the concept of cathexis: the combining or connecting of unconscious drive energy with ideas, affects, fantasies, the balance between the three drives (libido included) in interactions between people, and the emotional drive of the three drives into parts of personality and parts of one's body (Wolman, 1973). Ideas show cathexis of memory traces, whereas emotions are processes that discharge energy, giving rise to feelings in an individual.

A cathexis may be withdrawn from a preconscious idea, but the cathexis is retained in the unconscious. Freud also asserts the concept of anticathexis, as when cathexis is withdrawn from the preconscious back into the unconscious. This is an instance of primal repression and also an instance of construdo disconnecting from an idea of affect. Freud uses the concept frequently, in different ways, but he does not define the energy of cathexis, other than it is the mobile energy of the nervous system. Cathectic intensities in the unconscious are much more mobile. The whole cathexis of several ideas may be appropriated by the process of condensation.

Freud is describing processes that move from one place to another, that smash into each other creating new energies, which annihilate each other and then give rise to new energies, or new cathexis, without rhyme or reason in terms of conscious logic. We now know that subatomic particles smash into each other, annihilating each other and giving rise to the creation of new subatomic particles (Chester, 1978; Davies, 1982). In similar terms, a patient who was a successful writer said that he had "creative explosions," and then the ideas just flowed from his pen.

In particle pair annihilation, a particle smashes into its antiparticle, as when an electron and positron annihilate each other, producing a photon (Davies, 1982). Similarly, Freud (1900/1953a) found unconscious thoughts and affects were "transformed" into very different ideas and affects in dream images.

Freud did not explain but rather merely described the whimsical bouncing around of cathected energy. His discovery of these unconscious energies and their operation, however, is the first ever presented to the world. As they engage in a process of annihilation and recreation, subatomic particles are responsible for the unconscious mind's whimsical creations. These may move into the conscious mind.

It is the energy of construdo that underlies the tendency and direction of combining and connecting mental processes. These energies operate in accord with how Freud described the operations of an unconscious energy cathexis of a physical drive.

RETURNING TO UNCONSCIOUS DEVELOPMENT
OF CONSTRUDO OF THE FETUS

The fetus is combined and connected with its mother's uterus and blood supply. The folds of the embryo's and fetus's body are in direct contact with each other and with the walls of the uterus. This contact, or touching of body surface folds, fulfills the aims of construdo to combine and connect with its surroundings. The sensorium senses connection. There is also in the fetus's rudimentary psyche a biological striving to remain combined and connected with itself, to avert the greatest danger to the organism at this time – namely, a spontaneous abortion.

The fetus's psychic imagery is, according to TDT, being fired by the forces of physics in the universe. While the fetus exists within the surroundings of the mother's uterus, again, according to TDT, its psyche is being stimulated by electromagnetic, strong, and weak nuclear forces in the atoms of its developing brain, producing hallucinatory images of its connections and place in its surroundings. Concurrently, gravity also operates on this psyche in images of the psyche's position in its surroun-

50

dings. In this sense, the fetus's unconscious psyche senses how it is held in utero. Freud (1911/1958b) described such hallucinatory images as flashing through the infant's brain after its birth.

Construdo combines and connects the fetus in two ways: to the environment of its mother's uterus and to the forces of the universe that stimulate its rudimentary brain with hallucinatory images of these forces in operation. The latter (the forces of the universe) is a part of the beginning of creation of the unconscious psyche. The former (the mother's uterus) is a part of construdo's self-combining first stage of development. As earlier discussed, after birth and after about six months of age, construdo development leads the infant connecting with its conscious world.

Prior to this time, hallucinatory imagery and dream imagery have given rise to unconsciously determined images in dreams, in which things change from being firm under foot to being unstable and collapsing. Images of things are short and confined, then grow large and extensive. Things come at the dreaming infant, and then recede. Things get smaller, then larger. Such dreams are unstable, uncertain, because things keep changing. In TDT, this imagery is created by the forces of the subatomic particles. It is the environment seen from the unconscious mind during the first six months of life.

This imagery is the environment seen from the unconscious mind that the newborn tries to escape and get away from because it is so unstable and incomprehensible. The imagery is unpleasant because it transforms itself before the infant's eyes, similar to the way physicists report the particles being transmuted into different particles while being observed. The imagery is unpleasant and incomprehensible to the developing construdo drive's striving to combine the infant with what surrounds the infant. There is no definitiveness about the surroundings. The perceived environment is fraught with uncertainty. It is a perceived environment organized by what physicists call the uncertainty principle. Getting away from such an environment involves repression of such unpleasant perceptions, and combining construdo with a conscious stable reality, one that does not change. The infant's perceptions learn this about the surrounding reality.

Freud (1915/1957b) discussed the attention of the preconscious, a latently conscious zone generally capable of becoming conscious. The preconscious can also show censorship and repression of ideation pushing it back into the unconscious. It operates as a barrier, inhibiting the cathexis of certain unconscious ideation from discharge, or gaining consciousness. The attention referred to by Freud is also the construdo's drive aim to combine and connect with the surroundings.

The events of human lives in the environment, added to what they have experienced in reality, gives them the first telling of nature of the surrounding reality. Construdo is driving us to join our world. The derivatives of construdo can gain access into consciousness when they are so remote from the drive's original aim that they slip past the censorship of the preconscious, a process described by Freud (1915/1957b).

Humans form interactions with other humans from the earliest age. These interactions are the result of construdo. Construdo also drives individuals to combine parts of themselves with other parts of their mentality. It might turn an interest in stamps into having a stamp collection. Construdo also drives individuals to disconnect with parts of their mentality, as when they lose interest in a hobby they once had. The drive operates both consciously and unconsciously.

THE NEWBORN AND THE COMBINING STAGE
(FIRST SIX MONTHS AFTER BIRTH)

Consider the newborn and its interpersonal relationship with caring adults. The adults hold the infant in their arms, against their warm bodies; they stroke it gently with their hands, they rock it back and forth in their arms to cause the newborn to sleep. These acts involve physical bodily contact or physical combining. It is the first stage of construdo after birth.

Moreover, adults make physical contact with the newborn's exterior and interior mouth, by inserting into its mouth the nipple of the breast or bottle. They wipe clean the infant's anal sphincter and its genitalia. Again they physically combine with the infant; in these instances they approach entering internal openings in the newborn's body. None of the infant's body had been touched by outside sources before this time. These contacts after birth further serve to awaken the newborn's sensing of its own body, as the mouth, the anal sphincter, and genitals have not been touched before birth. We first sense a surrounding environment through physical contact, and such touch fulfills the aim of construdo's first stage – namely, that we combine with our surroundings.

Such an aim is unconscious, only discernable in evidence that all of these actions of the caring adult satisfy and leave the newborn infant contented. These actions soon become derivatives of the construdo drive's quest for satisfaction and release in combining. The aim of the drive's energy remains unconscious.

Earlier in utero, the developing embryo and fetus had bodily contact or physical combining only through touch with its own body. Therefore, self-construdo (or self-combining) precedes object construdo (or object combining). In either instance, combining is the first aim of the construdo drive, an intrapsychic drive whose purpose is outside the awareness of its participants. It can be inferred that we combine with ourselves, with objects, with our surroundings by touch, or physical contact, before we come to combine with our world visually.

Freud's (1905/1953b) discoveries of the libido drive and its component aims of oral, anal, and genital release of drive tensions can be seen as partly based on early physical contact with these specific bodily erotogenic zones by caring adults wiping and cleaning these areas. Construdo responds to the penetrations into these internal areas of the mouth, anus, and the genitals. They have been stimulated, as contacted combining areas. Freud also noted that the skin could become an erotogenic zone. We have already indicated how contact with the skin surfaces in utero and after birth satisfies construdo. All of this will in the adult be unconscious as component aims of these two drives, libido and construdo, that will be combined by the drive energy of construdo. Freud observed that the oral, anal, and genital erotogenic zone could be employed auto-erotically. Autoerotic behavior is the result of libido being combined with self-construdo. In its beginnings the libido drive was aimed at combining the embryo and the fetus within itself, via physical contact, as in the folds of the embryo and fetus giving physical contact against itself.

Discouragement of a child's drive to join or combine with its surroundings (e.g., by calling the child "dumb" or "stupid" or becoming hostile when the child attempts to understand its surroundings, and make new connections with it), can arouse anxiety and unpleasantness, which will result in the child's beginning to withdraw from any attempt to connect with its surroundings. Such actions diminish the child's intelligence. Construdo may disconnect or reverse its direction of energy toward objects in the surroundings that may have led to this unpleasant state of affairs. Construdo may regress to its first stage of self-combining, before connecting with objects or self as was its aim. This action disrupts or interferes with the individual's intelligence.

The frustration of a denied anticipated construdo satisfaction leads to the arousal of destrudo. Yet this denied anticipated construdo satisfaction may also be a most regressed elementary stage of the destrudo drive related to the thrasher aim, because the regressed construdo pulls destrudo with it. The regressed construdo reinforces destrudo's direction backwards. The thrasher's aim is basically to expend angry energy in all directions, lashing out with flailing hands and arms, feet and legs; screaming, crying, twisting its body in an expression of its (thrasher) destrudo's energies. The target of these angry energies is always undefined.

CONSTRUDO LEADS THE INFANT TO REALITY

Freud (1900/1953a) theorized that the infant during the earliest times is governed by unconscious or primary process thinking. Such thinking is hallucinatory in its efforts to achieve satisfaction or pleasure in the first six months of life. Then "a momentous step" occurs in the infant's psyche, when the infant comes to seek a real conception of its external reality. Now, in addition to the pleasure principle, the reality principle comes into play. This is a new, second principle of mental functioning in Freud's (1911/1958) thinking.

Concomitantly, the second stage of destrudo (pincer destrudo) is initiated when the infant attempts to gain control of surrounding objects it sees in reality. The infant takes hold of these objects between its thumb and forefinger and tightens its grasp. Moreover, this new relation with reality is augmented by construdo's new aim (the second stage of construdo) to connect oneself with external objects.

The construdo drive's aim (second stage) to form connections with the surroundings results in the phenomenon we call the reality principle. Freud did not explain it; he described it. When the infant's sensorium can well perceive its environment, the forces of physics drive it to construdo, to connect with the environment. Now it can focus its eyes on things around it. It's the second stage of construdo. The same forces of physics operate on all living creatures on our planet and produce a similar drive in them. One-celled organisms, such as the amoeba, with no nervous system or brain, connect with their environment and other amoebas almost as soon as they achieve life. A wasp will dig a hole in its environment to protect itself. A spider will spin a web to connect itself with its environment as a means of gaining food. This is not instinct; it is, in TDT terms, construdo drive energy from the forces of physics causing this behavior.

Construdo creates the infant's need to first combine the contact sensations of itself (visual, auditory, olfactory, and taste) with its surroundings, as well as with hallucinatory images prompted by its surroundings. These needs are connected with impressions of the caring adults around the infant. As the infant's self-perceptions have been combined into a sensing of itself, it sorts out its self-perceptions from the perceptions it has of objects in its surroundings. These are not consistent or congruent with its self-sensations and perceptions. The infant tends to perceive whole objects by the time it is six months of age. Here the infant is driven by construdo to connect with these objects. Thus, the reality principle, or a connecting with what surrounds the infant, emerges.

The attainment of the reality principle separates construdo from libido and destrudo in the id – the unconscious. Construdo is the first drive to emerge from the id into consciousness. Construdo also divides within itself. Part of it combines the infant with a sense of itself. Following birth, it gives the first differentiating experiences of self versus surroundings. Here construdo combines the infant with its surrounding reality, as well as continuing the infant's sensations of itself, and these begin to differentiate themselves from the surroundings. It is these differentiating aims that usher in the reality principle. After this differentiating time of self and surrounding the second stage of construdo unfolds. Then, the nature of the infant's connections with reality becomes its main focus. What the infant sees and how it is connected with what it sees, smells, touches, hears, feels, tastes, all tell the infant how it is connected with reality.

As Freud's "momentous step" implies, characteristics of the interpersonal connections are implanted in the infant's psyche. Construdo is not directed toward reality; rather, it is directed from the unconscious toward combining the infant with whatever surrounds it. It is the unconscious psyche's translation of energy based on forces of physics in the psyche to combine with things that are sensed. First, there are aspects sensed in the self and this is intrapsychic. Next, there are things perceived around the self. The infant's psyche is the next forming concepts of a psychic connection with these things or objects. It continues its connections with itself, as well as with feelings or emotions generated in itself, about these connections. The infant's psyche begins to form mental concepts in fantasy of what these combinings and connections mean or how they should be understood. Such fantasies can also begin the infant's distorted notions of reality, or the distorted world of the later child, adolescent, or adult. Generally, however, the infant achieves congruence in perceptions with the world's reality. Warm touching, stroking, and hugging responses from the caring adults generally convey the feeling that the infant is wanted, loved, and even adored, as it has been delivered into a warm, welcoming, positive place. Opposite behaviors will signal the infant otherwise. Either result can be established by the sixth month.

INTRAPSYCHIC AND INTERPERSONAL FACTORS IN THE CONNECTING STAGES OF CONSTRUDO (SIX MONTHS TO ONE AND A HALF YEARS)

The infant and the caring adults now, after six months of age, have more significant, meaningful interactions. Construdo drives the infant to seek and direct its attention to these interactions. These interactions prompt the second, natural, emerging stage of construdo – connecting the infant with its surroundings.

55

The infant smiles when the surrounding adults smile or laugh with it. These adults feed the infant when it cries to indicate that it is hungry. They change its diapers when its cries indicate discomfort or unpleasure. They rock it to sleep when its crying indicates that it is sleepy. The caring adults understand the infant's communications much better by this time, and so too does the infant understand the adult's communications and reactions. There are better connections between them. They have been engaged in communication for six months.

Interpersonally, connections are what is sought by the infant of six months to one and a half years of age. These connections are driven by construdo to make such connections with what is outside themselves. They distinguish a male father, different from a female mother. Fathers are protective, offer security, and interact with them in terms of male strength, such as tossing them overhead with muscular strength. Mothers are supportive and nurturing of the child's efforts to join the surrounding world, due to their continuous contact. The infant is sensitive to whom it is connecting with.

The child connects with what it is to be fathered, and what it is to be mothered. The child is beginning to get a feeling and mental imagery of the nature of these connections intrapsychically. Construdo uses the advancing mental capacities to perceive more nuances in perception in the sensorium, to know what the connections could mean – that is, to have ideas in the psyche that come together as a sensed feeling or concept about these connections. The child passes through the separation and individuation conflicts and issues described by Mahler (1968). Among the construdo issues at this stage are about how connected to parents versus how indepen-dent of parents the child wants to be.

The construdo stages influence individuals throughout their lifetime. Thus, individuals are driven to connect with others in a variety of ways. Young children join clubs or Scouts; feel a part of their school class. Adolescents join teams, fraternities, and sororities; form organizations; join others by doing what they do in fads, fashions, dress, buying and living habits. Adults connect in similar ways.

Children feel connected with the home they live in, their city block or county farm; they feel connected with their relatives and much that surrounds them in their environment. Adults continue to have these feelings in choosing where they will live. Adults seek to be with people who think and feel as they do, often with mixed results – sometimes satisfactorily, and sometimes disastrously.

INTERPERSONAL AND INTRAPSYCHIC INFLUENCES OF THE CONSTRUCTING STAGE OF CONSTRUDO (ONE AND A HALF TO THREE YEARS)

From one and a half to three years of age, the child, having made a connection both with the parent of the same sex and the opposite sex, constructs a mental bond with these parents or caring adults. In the latter part of this stage the child knows its gender. The child's psyche now constructs fantasy characteristics and qualities of the bond between these adults and itself. Relationships with peers in adolescence tend to define us; in them, we arrive at identity definitions of ourselves. These influences are part of the combining and connecting of ourselves with others – that is, the previous aims of construdo. In childhood there is latency, and in adolescence we may construct (or we may disconnect with) the perceptions our peers have of us, and decide to whether or not to carry such roles into early adulthood.

When our concepts of ourselves differ from the concepts of what we should be as conceived by our parents or the important peer group, we may experience a pathological condition of an identity crisis. We have constructed in fantasy a plan of what we want to be. When these ideas do not conform to what is occurring, a conflict in the construdo drive occurs between what these individuals want to be and what their parents and peers perceive them to be.

In early adulthood, the roles of our connecting with the world that we intended for ourselves (construdo fantasized) will or will not materialize. What we have constructed and created in fantasy, as a derivative of our construdo drive for ourselves to join the world around us, will or will not be realized. Failure to realize our goals can begin a persuasive depressing factor in our psyches.

The construdo drive to combine, connect, and construct relationships with our fellows will result in endless connections with other people. Doing what others in society do makes us feel connected with them. We construct roles for ourselves that others notice, and in doing so, we feel that we have constructed in our lives what others have constructed. We are like them.

In middle age, our construdo fantasies relate to the roles we have constructed for ourselves. We evaluate how successful we have been in fulfilling these roles. If we have missed our mark, a middle age crisis or depression may follow. During our lifetime, we may have had other construdo fantasies for what we wanted to do in life. These may have been pushed aside or forgotten because we did not have time to realize them. But now in our middle years, we have one last chance to revive them and seek the fulfillment of the fantasies we left behind from earlier years. To return to such fantasies will preclude and prevent a midlife crisis. By old age, we may have passed through the painful experience of losing friends to death, relocation, or disuse of friendship construdo bonds. Disconnecting from efforts to construct further friendships can be the end result.

INTERPERSONAL AND INTRAPSYCHIC INFLUENCES OF THE CREATIVE STAGE OF CONSTRUDO (THREE AND A HALF TO SIX AND A HALF YEARS)

The aim of the creative stage of construdo puts elements of the surrounding reality into a new combination of unique elements – different from the way such elements have been combined or connected before. This applies to any field of endeavor. In a variety of human activities, one can put things together differently than anyone else has before. It is an assertive statement of oneself. Strains of the combining, connecting stages, uniting oneself with others, continue into the unconscious construction of the creative product. In creating, individuals feel that they're breaking away from what others in their given field have done before.

There are transformations in a drive's basic aim as the individual develops and matures. Just as subatomic particles are transmuted into new and different particles in the subatomic physics of the atom, so too in the unconscious, transformations of the drive energies occur. These transformations are what Freud found in the course of libido energy gaining consciousness from the unconscious (Freud, 1915/1957b).

The first stage aim of construdo is to combine the embryo and fetus in utero within itself. This stage ends at about six months of uterine existence. It is followed in utero by the aim to form connections with the mother's uterine surroundings from six to nine months. After birth, these two self aims (combining and connection) arerecapitulated as object aims. Clearly the aim of construdo to combine with itself obviously exists in the infant. From six months to one and a half years, the child follows out construdo's connecting stage aim of making connections between itself and its surroundings. From one and a half to three and a half years, the child's constructing construdo stage aim is to put things together to produce additional things. Each of these earlier three stage's aims (combining, connecting, and constructing) is refined in the direction of encompassing more and more particulars of the individual's self-reflections vis-à-vis the surrounding environment. The individual combines with these particulars. The cumulative aims from the original construdo drive to combine, allows the new, refined drive to gain consciousness without repression of the original construdo aim. It is just as Freud found with the libido drive (Freud, 1915/1957b).

The self-construdo influence is continued in the creative construdo stage where the individual retains a combination and connection with the self. In addition to these fundamental combining drive influences, the creative stage aim not only retains the connecting and the constructing aims of construdo. This creative stage aim is also connected with others, in terms of doing and yet being different from what they do and what they are like. Thus, all of the earlier construdo stage aims contribute to the later self-creative stage aim. During construdo's self-creative stage, individuals develop concepts and ideas about what they want to be as people, sometimes to the extreme in glorified terms, all the way to in degrading terms.

So, stage by stage, construdo progresses under the impact or stimulation of the surrounding reality. What the infant, and later the child, perceives is incorporated into new drive aims. The aims are tailored to account for new information about what is reality. It is with this changing of the surrounding reality that construdo aims at connecting. It is from this changing reality that the individual attempts to make constructions. Only when the individual has mastered the three earlier stages' aims (combining, connecting, constructing) can the individual undertake forming an original, unique construction of a product, that while focused on the proceeding aims, shows a return to the aims of connecting with self, as seen in the first six months of uterine existence, and during the first six months after birth.

This process of maturation reflects the drive energies of construdo from the unconscious to be timeless and spaceless, as Freud (1915/1957b) found in the operations of the unconscious and as is the nature of the operations of subatomic particles. Freud indicated that drive energies were not annihilated in the unconscious, but rather repressed. However, in physics it has been found that subatomic particles annihilate each other. The collision of a proton and an antiproton annihilates both. They have different electrical charges. A neutron, with no electrical charge, colliding with an antineutron, with no electrical charge, also produces annihilation. They are opposites in their magnetic fields. The annihilation creates photons. It is a different view of reality and antireality (Chester, 1978). These particles exist for a minute fraction in time, "one hundred-thousandth of a billion of a billionth of a second in duration" (Chester, 1978, p. 135).

The unconscious operations can shift to new and different ideas, affects, or emotions in such processes. A new idea can come into the unconscious from two old ideas, affects, or emotions, annihilating each other in a fraction of a second and creating a new idea, affect, or emotion. A new creation can occur when the energies from the three different drives (construdo, destrudo, and libido) collide into each other and their different energies produce new creations. Directing these energies into each other is both an unconscious and a conscious undertaking.

The new idea, affect, or emotion can alternate or oscillate in the unconscious. So too, when the individual's thinking or feelings are stirred up the energized drives can come into collision with each other. This can result in the annihilation of parts of the emotions or affects.

The occurrence of collision happens when an individual is pondering part of the problem consciously, while other parts of the problem are ruminating in the three drives' energies in the unconscious. The collision of parts of the drives can create completely new ideas and affects. The self-combining track ofconstrudo is aroused and employed in forming these ideas within oneself. The individual is combining with ideas and emotions from the drive. Assimilated in the unconscious, a creative idea is born.

The Oedipal Conflict

In the child's fantasies are new sexual fantasies that are constructed in terms of exclusive possession of the parent of the opposite sex. It is due to the fact and observation that this parent has qualities that the child does not have and wants to possess, but can do so by constructing a special bond with this parent. The opposite-sex parent's genitals are of special interest to the child, because they are so different from the child's. The child's perception and awareness of reality by this time has perceived that people of opposite sexes fall in love or marry.

The fusion of construdo and libido results in a stronger bond that either of the drives alone would create. Libido is connected with construdo to construct an unassailable unity, or bond. This fusion and bond dictates doing away with the competition (i.e., the parent of the same sex that the child does not want to be there). The presence of this same sex parent arouses the child's destrudo.

The child's manner of connecting with the world that surrounds it, and its own constructing of reality, may make the same-sex parent feel overlooked, ignored, or irrelevant. The parent may become hostile and attack the child's construdo's development. The child experiences such reactions as discouraging its efforts to make connections and to construct for itself its own desired reality. In such cases the child's intellect and construdo efforts to learn and connect are being attacked. The child overgeneralizes or overconnects the idea from its drive motive to totally possess the parent of the opposite sex. The child may withdraw this part of its construdo, further connecting with its parents in a way that possesses one and eliminates the other; this can result in the child's reluctance to connect with reality to varying degrees. It resists learning about the world (learning equals connecting in the unconscious). Therefore, at this time and beyond the child may find it difficult or be unable to learn.

The child of three and a half to five years of age has constructed in its psyche and its fantasies a relationship bonding it to its parents. The relationship is a result of construdo's combining, connecting, constructing, and creating influences beyond libido.

After three and a half years of age, construdo moves into its creative stage. The aim here is to construct something from the psychic elements that have been combined, connected, and constructed intra-psychically. By this time the child has constructed in the psyche a concept of what the bond or relationship between its parents and itself is. How the child has conceptualized this relationship is crucial to the Oedipal conflicts created and how these conflicts will be resolved.

Libido is aroused in the child as it enters the genital stage of this drive (Freud, 1905/1953b) (construdo) and seeks release of the tension built up in the genitals. The object of libido becomes the parent of the opposite sex. Now a rivalry with the parent of the same sex for the parent of the opposite sex begins. It is a triangular conflict. It is thought that in such cases the frustration of the child's wishes or fantasies to possess the parent of the opposite sex arouses the child's destrudo. With the arousal of destrudo and the child's fantasy of removing the perceived competing parent, a conflict arises, because the child also has a strong construdo constructed bond with the same-sex parent.

This situation stresses the child's capacities to resolve conflict within itself. Often, the child is not able to resolve these conflicts, which leads to continued pathology in the child's object relationships; later objects in the child's life may become substitutes for the objects in the original Oedipal conflicting triangle of objects and drives. These multiple bonds underlie the powerful effects on the psyche in the Oedipal conflict observed by Freud (1905/1953b). It is the first time all three drives (construdo, destrudo, and libido) have been cast into such intense conflict. The child may have a poor idea about how to resolve the conflict or no idea at all. Unresolved, the conflict persists in the child's mind. Later relationship with the parents may be affected adversely.

Love

Both love as one track of the creative stage's aims to combine and connect with others, together with the constructing stage aim, forms something between the two individuals and leads to original, unique ways of seeing each other; each creating ideas and affects about the other, while others around them do not always see what they have created in their minds about each other. Thus it can be said that in one way love is truly blind.

Combining, connecting, constructing, and creating are the underlying construdo aims behind love. The love goal is so remote from these unconscious energies that it is not understood as a derivative of drive energies when it comes into consciousness. When construdo coupled with libido aims toward individuals, the two drives in tandem produce even greater energy. Partners in the couple, in their creative thinking and emotions, idealize each other. The extent of their creativity defines the depths of the love created between them. It may be connected with all the ideas in their construdo history they thought and felt love should be in their lives. They bring all these fantasies into their newfound love.

When the parties view the connected, constructed relationship in very different ways and feel let down by the other, they may be disappointed to find that they have very different construdo expectations of how the relationship should proceed. Destrudo can arise between them. In these circumstances, the love can change to hate. Love alternates here between belief and disbelief through the time that the love fantasy was true or not true.

Giving up a love affair or disconnecting ourselves from an all-encompassing love relationship is usually soul wrenching and painful. It takes time to work it through mentally, because one's self-creative construdo (who we are in the love connection) and our created fantasies (of who the other person is require us to give up and to dismantle cherished construdo fantasy parts of ourselves) – are our very own creation. It is painful to dismantle our own created love bond. It takes time to convince ourselves that the fantasies we believed no longer exist. They are destroyed or smashed by the love partner's antagonistic, hostile, nonsupportive behavior toward us. We begin to feel the same destrudo feelings toward them.

Ultimately, we disconnect from the object or loved one because the destrudo parts of us, which have quietly alternated with the construdo parts of us in the past toward that person, now dominate our thoughts. We become disconnected from our drives toward the person and not drive connected with the person. Sometimes construdo allows a friendship to continue between the two. The failed love relationship can change into a friendship because the construdo bonds outlast the destrudo disconnections, perhaps just as the uncertainty principle in physics led to uncertain observations of a particle in subatomic operations.

A Note on Creative Construdo

When self-creative construdo in thought, emotions, or self-sensings is combined and connected with object-creative construdo along some particular or specific line (as self and object destrudo in oneself), and travels through the unconscious, these original ideas can produce macabre, object-creative construdo works of art in the conscious productiveness of individuals. The paintings of hell by the 16th century Dutch artist Hieronymus Bosch and the frightening stories of the 19th century American writer Edgar Allen Poe bear witness to the point.

CHAPTER 5

CONCEPTS OF THE DESTRUCTIVE DRIVE

PSYCHOLOGIES AND THEORIES OF THE DESTRUCTIVE DRIVE CONTRASTED WITH DESTRUDO

Freud (1920/1955a) introduced his concept of the destructive drive in *Beyond the Pleasure Principle.* In this work, he first asserted the death instinct. Further, Freud asserted that the musculature of human beings was the executor of the destructive drive. Triadic Drive Theory (TDT) believes the musculature is the executor of the drive in its formulation of destrudo. Freud asserted that the drive's aim was backward in time. Again, TDT concurs.

Freud never presented stages of the drive's development or their aims, as he had done with the sexual drive, libido. TDT has sorted out the different stages of aggressive development and the different aims of each stage, as they follow the developmental muscular capabilities of the infant and child. (These are given in Chapter 4.) It is crucial to understand these influences in a variety of destrudo situations.

Freud did not present neurotic symptoms, character traits, or psychotic symptoms that might result from the twisted, out of balance ramifications of destrudo, as when it operated without mental controls. TDT's theory of the unconscious aims of the drive remains consistent with Freud's observations of drive energies stemming from the unconscious and gaining entrance into consciousness as unconscious derivatives of the drive. It is the remoteness of these derivatives from the original aim of the drive that allows their entrance into consciousness (Freud, 1915/1957b).

However, TDT has asserted that destrudo unconscious energies act like the energies of subatomic particles within the atom. Freud's understanding of the unconscious operations is congruent with our understanding of subatomic particles' operations in the atom. In fact, it seems that Freud's description of unconscious operations foreshadowed knowledge of the operations of subatomic particles. They are parallel because the energies of physics in the atom cause the operations of the unconscious in the human psyche. Annihilation of subatomic particles and the emerging of new particles is one origin of the destrudo drive in human beings. The original influence on destrudo outside the human being is from the Big Bang at the beginning of the universe. We will see that this influence is registered in the unconscious.

According to Freud, time and space have no meaning in the unconscious, nor do they have any mechanistic meaning in subatomic particle physics. An emotion, a drive's energy, or an idea can move from one place to another in the unconscious. This is called a displacement. An emotion, a drive's energy, or an idea can come out of the unconscious into consciousness, transformed into a somewhat different emotion, energy, or idea.

As stated throughout, TDT departs from Freud's thinking that the drives are biological and asserts that destrudo and construdo stem from physics and its energies; only libido is a biological drive. Freud (1920/1955a, 1930/1961)saw life instincts, his libido drive (Eros), as opposing or pitted against the death drive (Thanatos). TDT believes construdo, not libido, is responsible for maintaining life. Libido reproduces or recreates life.

Freud (1920/1955a) stated, "The attributes of life were at some time evoked in inanimate matter by the action of a force of whose nature we can form no conception" (p. 38). TDT asserts that the force that Freud could not conceptualize was that of the binding forces of physics. Freud (1920/1955a) said further that these instances of creating life perhaps succeeded and failed many times before life came to sustain itself. He was thinking about how the force got life started on our planet and what resulted in its continuing to succeed until it took hold. He referred to our sun as reinforcing life, suggesting his awareness of our overall universe's forces in these matters.

Just as Freud found that repression of libido leads to mental illness, he also explored the question of mental illness with regard to destrudo. However, he arrived at different questions and different conclusions. By 1915, Freud had differentiated between the sexual instincts and the ego instincts. Eros and the ego instinct "seeks to force together and hold together the portions of the living substance" (Freud, 1920/1955a, footnote, p. 60). Precisely this function encompasses the concept of construdo. This force, Eros, operates from life's beginning as a "life instinct" (Freud, 1920/1955a, footnote, p. 61). It comes into being when life comes into inorganic substances.

TDT has asserted that this function is of a separate, distinct drive, namely construdo, and that the force behind its functioning stems from energies of the binding forces of physics. This is a clearer, more precise, more defined description of what Freud ascribed to his extended concept of libido and narcissistic libido in his life instinct. Again, TDT holds libido to be the drive to recreate the species. It is the only biological drive. This contrasts with construdo, which is the drive from physics that builds up organic substances and mental ideation into larger (and even larger) combined units. It is contrasted with destrudo, the drive that breaks these things apart.

Freud (1920/1955a) said the transformations of the concept of the ego instincts first were distinguished from the sexual instincts directed at an object. He next considered a libidinal aspect of these instincts directed toward the self in narcissism and self-preservation. He recombined them into libido. Thus sex becomes the most important part of his life drive. Yet, finding an opposition in the psyche to this combination, Freud was led to assert the death instinct in opposition to his combined life instinct (Freud, 1920/1955a, footnote, p. 61). Freud then saw these operations of the ego as being mostly unconscious. Construdo and destrudo's aims and influences are mostly unconscious.

Freud's thinking about energy from the unconscious psyche for the destructive drive follows his concept of reduction of the internal tension arising from the death drive. His constancy principle is based on mechanistic thinking of Newtonian physics before the era of the quantum-mechanical model – that is, after the uncertainty principle in physics, knowledge of the operations of subatomic particles had to include the probabilities of quantum theory. Freud's (1900/1953a) perceptions and intuitions were predictive and accurate of truths of nature yet to be discovered.

Freud (1920/1955a) decided that the pleasure principle did not hold in the matter of the death instinct; hence his title, *Beyond the Pleasure Principle*. However, when construdo unites destrudo with other mental factors or concepts, a wide variety of expressions can result. These expressions give satisfaction to the construdo drive, and its aims to make connections between mental events, and construct new mental concepts in the combinations. There is a multiplicity of destrudo expressions that construdo can connect against ourselves or against others outside of us. These expressions are outside any fusion with libido, as in the case in Freud's concepts of masochism and sadism.

Lorenz (1966), in *On Aggression*, indicated that aggression is a biological factor, an instinct beyond the control of human beings. Lorenz said aggression can erupt unexpectedly. Aggressive energy accumulates in the nervous system, and when not released, it can explode. TDT defines many disguised ways that aggression is released against others or the self, but Lorenz's idea is somewhat modified in TDT. Aggression, he stated, comes from energy inside the human being, and not in reaction to outside stimulation. Outside stimulation need not be awaited for release of aggressive energy. TDT concurs. Aggression is more than a type of reaction to outside stimulation or provocation. It is a psychic energy from inside the person.

The psychological creations of human beings often hold the capacity to stimulate and release aggression, but in the absence of such psychological creations, aggressive energies will release themselves with little provocation. Lorenz's (1966) observations are due to the fact that destrudo is in part determined by uncertain operations of subatomic particles in the individual's brain. According to Lorenz, aggression serves the purpose of survival of the fittest, self-preservation, or survival of the species. This follows, as Lorenz sees it, if aggression is considered to be a biological drive. Lorenz (1966) also asserted a value judgment regarding aggression, seeing it as an evil of human beings. That is, it is an instinct that has become exaggerated in human beings, beyond its original purpose for survival. In contrast, TDT believes aggression or destrudo is more than a drive to defend us for survival. It is a natural part of us, often disguised to gain acceptability in the conscious world. It seeks justification, but it influences us whether justified or not. In many instances it is not justified in terms of the requirements of the conscious world. It can come into conscious expression one way or another – justified, rational, or not. Often it is rationalized in the most illogical ways. Its justification can become a gross distortion of reality.

Lorenz (1966) saw that human beings need sufficient outlets for their aggressive drive. They suffer from civilization curbing this drive. Human beings often express their aggression in ways that cannot seemingly be comprehended. Moreover, when individuals do not express their aggression, they release the drive against themselves in disguised ways. Therein lays the real danger from the drive.

TDT believes the drive is released in one way or another. When released against the self, people cause their minds to create their own sickness, without having any conscious awareness of it. TDT postulates that therefore individuals can kill themselves in the process. This process is what Freud saw as the death instinct. His observation is that the meaning of human behavior gets off track. To put it back on track, we need only say that if such behavior is allowed to go on uncomprehended, it will look like the individuals are wishing to kill themselves. This is not the case.

Destrudo theory does not consider aggression as an instinct or an innate drive for survival or self-preservation. Rather, it is the result of forces of physics that would break all human existence apart, as in a collision of two subatomic particles in our brain, or as in the Big Bang that initiated the universe. We consider aggression or destrudo to be a drive continuously occurring in the unconscious from the continuous operations in the atoms of our brains, from the Big Bang 13.8 billion years earlier and the destructive universal energies surrounding us, constantly being registered in our unconscious. Since the unconscious and its forces are timeless, it does not matter how far back in time the aim of the drive originated. It continues to this day. Its outward aim is ultimately to break objects apart, so that they can never exist in reality again – to destroy them completely.

The fact that human beings have used their aggression as the means of self-preservation and survival does not establish that this is the reason for its existence. Lorenz (1966) is thinking of the drive as biological. It is the same aim that Freud (1915/1957a) asserted in "The Instincts and Their Vicissitudes." Freud (1920/1955a) also asserted that there was no pleasure in the satisfaction of the drive's aim (i.e., to go backwards in time, to an earlier state of the inorganic, or to what he saw as the individual's death). In TDT there is the pleasure of having the construdo drive's aim fulfilled. The destrudo aims are thus connected with certain situations rather than thinking the situation caused the destrudo reaction. Construdo connected destrudo aims to the situation. This entire process can be understood without resorting to biology for explanation.

The frustration-aggression theory of Dollard, Doob, Miller, Mowrer, and Sears (1939) asserts that the frustration in the lives of human beings will lead to one or another type of aggression. In general, TDT agrees. But because this theory assumes aggression is based on the biological nature of individuals, its origin has little relationship to destrudo theory.

Differing from TDT's analysis of destrudo is that of Erich Fromm (1973) in *The Anatomy of Human Destructiveness*. Fromm's approach is a sociological, biological, and historical analysis of human aggression and destructiveness. He asserted that there are two completely different types of aggression. One, shared with the animal kingdom, is the impulse to fight, attack, or flee when threatened. It is a biologically adaptive reaction for survival. When the threat ends, the impulse subsides. It is "benign" aggression. The second Fromm called "malignant" aggression. It involves destructiveness and cruelty for its own sake. Such aggression is found only in human beings. Previous thoughts on aggression have failed to make these distinctions. The two have different sources and different qualities. Fromm (1973) noted that destructiveness increased as civilization developed. Humans are killers. They kill and torture their own species. They are the only primates who enjoy doing so. It shows the malignant nature of their aggression. He saw the preoccupation with decay, decom-position, and death (necrophilia) as evidence of malignant thinking. His view makes value judgments. It is our view that the malignant thinking shows that these individuals are linked with the decay and decomposition facets of destrudo's progression toward breaking a dead body apart into pieces. Fromm (1973) saw the difference between the two types of aggres-sion as fundamentally based on the difference between an innate, or organic aggressive drive, shared with animals, and a character trait that exists in human beings alone. Fromm rejected Lorenz's (1966) idea of a transformation of beneficial, defensive aggression into destructive cruelty as a progression in human beings. Animals, he noted, are not torturers or killers.

Fromm (1973) saw human nature as characterized by fundamental contradictions. He attributes these contradictions to the biological dichotomy between diminishing instinctual behaviors in human beings and growing self-awareness. He noted that humans are the only animals that do not feel at home in nature. Humans overcome their separateness, powerlessness, and sense of being lost by seeking new ways of relating to their world. Human separation from their instincts produces a conflict of being alienated from the nature of their existence. Yet the expression of their aggressive instinct is one of the means through which humans can stay related to themselves. However, TDT attributes this expression to construdo's promoting individuals to combine and connect with others using their destrudo against them. Fromm's (1973) analysis begins from outside the individual. He looks at a variety of human aggressive behaviors. He looks at roles sought out by individuals in society and their historical roots as influences in causing certain aggressive character structures. He works his way inward in his analysis toward explaining their inner aggressive character structures. His analysis involves more judgment of the aggressive behavior, rather than explaining the roots of the behavior.

TDT's analysis of destrudo starts from inside the individual, citing the operations registered in their brains and in the atoms of their brains. When a subatomic particle smashes into a subatomic antiparticle the result is particle annihilation and the creation of new particles. Prior to this, TDT postulates that there's the timeless beginning of destrudo from the Big Bang, 13.8 billion years earlier, registered genetically in one's unconscious. Last, there's the influence of countless destructive happenings on the unconscious from the surrounding universe, such as galaxies colliding into each other and the resulting explosions as stars in these galaxies collide. There is a destructive energy influence stemming from the comets and asteroids smashing into stars, planets, and satellites in our galaxy and all other galaxies. It's the registered unconscious brain influence of destructive energy from nature's disasters on our planet, such as volcanic eruptions, earthquakes, mudslides, floods, tornadoes, hurricanes, and violent snow and rain storms. Energies absorbed in us from such happenings augment the building of the destructive drive, destrudo, in us. From all of these three sources (operation of atoms in the brain, particle smashings, and destructive natural disasters), the destrudo drive originates.

CHAPTER 6

THE SEVEN DEVELOPMENTAL STAGES OF SELF
AND OBJECT DESTRUDO

THE CONCEPT OF DESTRUDO DEVELOPMENT

Although writers in psychoanalysis have observed great variations in aggressive behavior and its aims (Brenner, 1971; Hartmann et al., 1949; Stone, 1971), they have never agreed on the definition of a unified destructive drive. This chapter presents a formulation that distinguishes the types of aggression and their aims, proposing a theory that specifies the origin of this behavior. Understanding of destrudo behavior will be augmented by an outline of the different developmental stages of destrudo and their divergent aims.

In utero there are two forms of self-destrudo development: one transitional stage after birth from self-destrudo to object destrudo, and then four stages of object destrudo after the sixth month of life. This is a conclusion based on the capability of the body's musculature to perform certain destrudo behaviors. At certain phases in its development, the body can carry out new operations because of new capacities that evolve through growth and maturation. The musculature can then perform new actions that have destructive potential and are used to accomplish destrudo goals, as well as the other aggressive actions.

From the beginning, destrudo energies in the archaic unconscious combine with musculature actions in the archaic psyche to form its destrudo portion. This effect occurs after birth. Freud (1920/1955a) conceived that the human musculature was the executor of the destructive drive. Ever since then, psychoanalysis has been unable to move past Freud's declaration. Let us now consider an exposition of the stages of self-destrudo.

Break-Apart Self-Destrudo
(First Six Months in Utero)

In the first two trimesters of pregnancy, the spontaneous abortion of the embryo/fetus may occur from external injury, disease, or defective genetic development, or when the mother's hormonal balance is insufficient to continue the pregnancy. When the embryo/fetus dies and begins to decompose, it is expelled by muscular action of the mother's uterus. If this process of breaking apart leaves some parts of the embryo or fetus in the

71

uterus, these leftover parts must be removed by a surgical dilation and curettage.

This results in destruction of the organism, as what has been achieved through growth is broken apart in a backward progression to parts that had previously grown together. Therefore, established destructive processes in the organism in utero unfold just as surely as normal forward developmental processes unfold in utero. From this it can be inferred that both processes are potential outcomes for the human organism. Life can advance forward or retreat. Clearly, the backward destrudo process is in evidence in every spontaneous abortion.

The physiological, biophysical, and biochemical factors responsible for this outcome are outside the scope of this exposition because the backward process is our primary concern. Death does not end the process, as Freud thought, but is only a phase in processes that continue to break the body into pieces, followed by organic decomposition, then a breaking down into inorganic atoms. The opposite of life is a retreat to a time before life existed. Triadic Drive Theory (TDT) calls this state "absolute death."

The backward and forward directions in the development of destrudo and construdo appear to originate in the primitive developing brain (lower brain stem) of the embryo and the fetus. This same part of the brain of a pregnant woman also appears to govern the muscular contractions that cause fetal expulsion from the uterus. All these processes of physics are recorded in the developing, archaic, unconscious psyche.

The aim, then, of the first stage of self-destrudo development is to break apart what has been assembled imperfectly in forward human development and to reverse the forward advance into backward progress-sion. This first stage of self-destrudo seeks to take this reversed direction of life back to a time before any life existed, when there were only the atoms of carbon and oxygen formed which humans developed when the planet was a mass of cooling gaseous elements – a time before any organic life was yet to occur. This one archaic, unconscious aim of self-destrudo is first implanted in the developing unconscious psyche in the utero. TDT believes it is this archaic unconscious aim that underlies the so-called nirvana principle (Freud, 1924/1961).

To eliminate all energy, bring the life processes to an end, and reduce them to an inorganic state where the energies attain perfect equilibrium or a merger with the cosmos would require a return precisely to the time and state just described as the ultimate aim of destrudo. This would be a return to when hydrogen and helium atoms in an even earlier time and state of the universe following the Big Bang (an earlier aim of self-destrudo) had cooled to the point where they converted into the carbon and oxygen atoms from which all humans are made (Hawking, 1988).

The ultimate aim of destrudo to destroy an object begins to influence the organism when it is first forming in utero. The human organism comes together under the impact of construdo created by the binding forces of physics. The external influence of these forces also creates destrudo in living organisms and in their psyches. The directions of destrudo toward backward progression or of construdo toward forward progression in organisms are the result of different forces of physics interacting in the human organism from its earliest formation in utero. The embryo/fetus senses the forces that could break apart its existence and strives to overcome them by supporting its construdo strivings to continue life.

In TDT thinking, psychological factors can influence these stri-vings that occur between advancing the existence of the forming organisms or by reversing the combining process in the organism and breaking it into pieces. It is the mother's feelings toward her pregnancy that create these fetal psychological factors. If she desires her pregnancy and is pleased by it, then the psychological factors, the sensory communication between mother and embryo/fetus, are favorable for its continued forward development. On the other hand, the mother may have mixed or negative feelings toward her pregnancy; she may fear childbirth, or feel her life might be negatively affected by the pregnancy or the birth, or even believe it would end all dreams she had for her own life. In that case, the psychological factors in the embryo/fetus for sensing positive emotions about its very existence, development, and progression become very unfavorable during the fetus's first six months in utero. It is proposed in TDT that this can have life-long negative effects.

Some external injury may cause the child within her to struggle to sustain its life by invigorating its construdo. In these instances, the fetus encounters a negative uterine environment, instead of a negative maternal attitude toward the pregnancy. The effect on the fetus is the same, as its continued survival depends on its own construdo fortifying it against outside threats. Construdo is the drive to preserve life.

Sometimes the mother's career is threatened by the pregnancy, or she may feel unhappily tied to her partner by it. She may envision rearing the child alone because she is not truly aligned with the father or not married to him. Economically, she may have no way to support herself and her child. Her pregnancy also means she will have to assume the role of a responsible adult; she may feel unprepared for this role, because she feels like a child herself.

TDT proposes that these and other negative feelings on the mother's part can be sensed by the embryo/fetus, causing self-destructive motives. Besides the physiological, biophysical, biochemical, and accidental dangers to the in utero existence of the embryo/fetus, the psychological feelings can be adopted toward itself, creating urges of self-destruction or self-destrudo. Whether to survive uterine existence or to die is only sensed in the psyche of the fetus's forming brain and psyche. This mortal choice is the first conflict of the psyche every human being experiences in its primitive psyche.

In psychotic persons who are governed by their unconscious, the comings and goings of these component motives can be observed and assumed in direct accord with the operations of subatomic particles that come and go, changing their identity.

Construdo is the force for survival in utero, whereas destrudo is the force for death or the end of the fetus. The continuation of self-destrudo motives in psychotic individuals can be observed when they jump from high places to smash their bodies, leap in front of oncoming trains or cars, or run their own cars into walls and smash themselves. The derived aim of self-destrudo at this stage is to smash the self to pieces. This is the break-apart self-destrudo aim.

Defective Part Causing Self-Destrudo
(Six to Nine Months in Utero)

By the third trimester, the human organism is completely formed and gaining mass. However, any of its internal organs may be defective due to genetics, a disease process, an external injury, or because the mother's hormonal balance does not facilitate the organ's proper development. A defective heart, liver, or kidney may be incapable of sustaining life; the result is almost always a stillbirth or the newborn's death soon after birth. One defective part can cause the death of the human organism; its forward development ends, and its backward recession begins. The destrudo aim at this point is to end forward development because part of the organism is defective.

In this backward progression (recession) there is death, decomposition, and ultimately (in hundreds of years) a breakdown into inorganic atoms, the state of things before life existed when there was absolute death. This is the aim of the second self-destrudo stage: defective part causing self-destrudo. This living or dying conflict is registered by the archaic unconscious psyche as a potentiality, reality, or possibility.

When life continues and the combining of living parts wins with the human organism surviving after birth, this struggle against the possibility of fetal death remains in the unconscious. A defective part of the developing fetus thus threatens its continuing life as a motive in the unconscious because the unconscious is illogical and permits logical contradictions. Any defective part in the fetus can implant in the unconscious a notion of self-destrudo because the whole organism could break down and die from a failure of this defective part.This example of self-destrudo is evidenced in depressive psychotic thinking when it is determined by the unconscious. These psychotic adults want to kill themselves because of some imperfection they perceive in their character, their history, or their ways of dealing with people. This is construdo combining the self with an unexpected imperfection, so that the destrudo is aroused, viewing the self from the unconscious perspective. Thus, self-destrudo motives are aroused in the unconscious and such individuals may want to do away with themselves. Sometimes they injure some part of themselves, without dying.

These self-destrudo acts can be construed as inflicting a death blow to some part of the self.Psychotics will put guns to their heads and pull the trigger, stab knives into their hearts, put nooses around their necks, and jump off chairs. Attempts like these are likely to succeed. Psychotics draw some self-destrudo unconscious aims from the previous self-destrudo stage of break-apart self-destrudo. Individuals whose mothers did not want their pregnancies may choose to slip out quietly by turning on the gas or taking poison, much like their existence in utero – quiet, undramatic, but self-negating. In summary, negative life experiences in utero can impact powerfully on the unconscious, causing severe problems later in life.

THE SELF AS OBJECT

The preceding two stages leading to self-destrudo are the result of construdo combining some part of the self with destrudo; but part of construdo also combines with self-connections in the unconscious. Following birth, during the first six months of life, self-combining and connecting sensing of construdo from uterine times continues in the human psyche's unconscious. Similarly, self-constructing and self-creating sensing in the human psyche's unconscious continues in the newborn. So too, the two self-destrudo stage aims from uterine times are held in the unconscious after birth by construdo influences.

Transition from Self to Object Destrudo

Hartmann (1964) asserted that newborns do not differentiate between self and objects. According to Hartmann, this perception of self and object develops out of a common matrix formed by id and ego. TDT believes this formulation also applies to the developing fetus and to the newborn until six months of age. The process that turns self-destrudo aims into object-destrudo aims is birth.

Thrasher Self/Object Destrudo
(Birth to Six Months)

At birth and during the following six months, newborns kick at the air with feet and legs, hitting the air with hands and arms. Destrudo has been mobilized, but what is its target? The newborns twist their necks and bodies. They cry and scream using the musculature of their vocal apparatus, venting feelings vocally but without language. Their musculature attacks – they know now what. During these actions newborns exhaust themselves from all this activity and eventually sleep. The baby has yet to differentiate its target as inside or outside itself, as it flails angrily with its limbs, twists its torso, screams, and cries. Psychoanalytic theorists and developmental psychologists agree on the baby's lack of object differentiation at this time (Hartmann, 1964; Jacobsen, 1964; Piaget, 1954).

At the moment of birth, the newborn reacts immediately to this radical change, lashing wildly at its new environment in all directions. As soon as it is delivered, the newborn indicates the intense displeasure it feels. This expression of destrudo is fitting indeed, because it validates destrudo's basic aim to turn events backward to what existed previously. Before birth the fetus knew only the uterine environment of warm fluid, darkness, silence, and a constant temperature. The uterine environment also provided close sensual contact with itself and the uterine walls, as well as a constant blood supply of nutriment and waste product elimination. The newborn has no effective way to achieve the reversal of going backward in its development. Birth is the first trauma and the first frustration of newborn life after it has been independent only for a few seconds.

Both Leboyer (1976) and Adler (1956) interpreted the newborn's furious reaction toward the progression into birth as being intensely aggressive. Or as Stone (1971) expressed it, there is an intense desire "to return to that relatively homogenous stable condition experienced in a total organismic sense before birth" (p. 236). The intense destrudo reaction expressed by the musculature of the newborn in these circumstances comes to an end only when it reaches the point of exhaustion and the newborn goes to sleep. Confirmation of this musculature activity is provided by Ilg and Ames (1955). Thrashing of the musculature typifies this period.

Destrudo has become combined with the developing musculature by the construdo operant need to combine things. The newborn is expressing its displeasure through destrudo. Gaddini (1972) declared that displeasure is linked to destrudo, and pleasure – to libido. Anna Freud

(1946) indicated that studies confirmed that an infant's experience of pleasure promotes libidinal growth; if it experiences unpleasure, that promotes destrudo. Yet these conclusions do not go far enough. When construdo joins in the mix, it can also promote pleasure; and if coupled with libido, so much the better. Because of its aims, destrudo causes tension, stress, and unpleasure. Triadic Drive Theory (TDT) continues Freud's (1920/1955a) concept that the musculature is the executor of destrudo. It further views the development of the infants' musculature as creating new potentials for destructive actions.

Bear in mind that these are phases of destrudo energy. In physics, energies show phases in flow with their different intensities, not sites where they are focused. In TDT, we find no biological body zones of focused excitation, such as Freud observed with the biological libido drive; that is, forces of physics in us reveal phases of energy, not zones of bodily excitation.

The infant wants food (milk), body contact (to be held against another's warm body), sleep (being rocked to sleep), clean diapers (having its wet or dirty diapers removed from its body). If it is cold, it wants to be made warm, by being held or covered by a blanket. Dissatisfactions such as these, if uncorrected, arouse the infant's destrudo.

During the first six months of life, an infant learns to differentiate itself from objects in its environment. After that, destrudo arousal is prompted by construdo's failed expectations of surrounding objects or of the self.

Destrudo behavior during the thrasher self-object destrudo stage (first six months of life) begins a pattern that is often returned to in childhood, such as during temper tantrums. The body expends energy in all directions, often accompanied by screaming, until the infant or the older person is exhausted.

A transformation into thrasher destrudo occurs after a person is born and has to adjust to a completely new environment. The unconscious is affected by environmental conscious-combining influences, such as when individuals are angry but do not release or express their anger. Instead, such individuals exert themselves energetically in several different directions until they are exhausted through muscular activity or by yelling and screaming. Before the infant has language, it lashes out at unspecified targets in every direction. It continues until, at some time during the first six months, the infant develops the notion that the target of its destrudo is some object amid the reality around it.

By six months of age, the development of a prehensile finger grasp and the ability of eye muscles to achieve binocular focus on objects with new found depth perception cause new specific connections with the infant's surroundings.

Self-destrudo aims that conflict with object-destrudo aims characterize the thrasher self-object stage. At birth the undifferentiated perceptions of the newborn are forced to distinguish between what belonged to self-perceptions and self-combining and object-combining within its surroundings. Gradually, the infant gives up these unconscious psyche perceptions of reality. These unconscious psyche perceptions of reality are too contradictory, changeable, illogical, and unpredictable to permit the infant to achieve the stable concept of the reality it senses around itself. How could the infant's mentality build logically, faced by such constantly changing perceptions? By six months the newborn concludes or senses it can't. So, it seeks a new set of reality circumstances and then the reality principle begins.

Construdo disconnects itself from adding its influence to such perceptions because it is too unpleasant, uncertain, and mentally disconcerting. Construdo does not seek to attach itself to such fluctuating realities. Rational, logical individual concepts develop by contrasting themselves with incongruities. The infant seeks to reduce and eliminate these contradictions in perceptions. The conflict between self and object perceptions continues.

When adults are fixated at this destrudo stage, destrudo can cause them to attack another to whom they feel close and simultaneously perceive that part of the attack as aimed at themselves. Patients with borderline personality disorder experience this effect. When they attack someone they love, they also feel bad and attack themselves. The confusion here about destrudo's target is from the stage of thrasher self/object destrudo. It can be activated when a patient's destrudo is aroused, because their destrudo is fixated or arrested at this stage. It stems from infancy; when frustrated, the infant frets and angrily thrashes the air with its limbs. It is angry at itself. But it also screams at the mother who is not making things better. Either can be the target of its anger, and at times both are. The infant attacks both the self and the object with whom it's linked. When this destrudo stage is reactivated in the adult with borderline personality disorder, both self and objects are attacked.

The same conflict can be aroused in adults who are beyond the borderline personality disorder level of personality organization. These adults cannot decide to direct their destrudo toward someone who has injured or hurt them, because they fear greater retaliation and further injury to themselves. So they hold back their aroused destrudo toward the provoking individual and do not release it. At the same time, they are angry at themselves for not carrying out what their destrudo drive is urging. Such mixed destrudo aims can reactivate the thrasher-destrudo stage aim from earlier times when this confusion first existed. Regression to this stage of destrudo toward an object that cannot be expressed can leave the self as object. The unexpressed destrudo can cause the self to absorb these destrudo energies against the self as in physical illnesses.

Pincer Object Destrudo
(Six Months to One and a Half Years)

The first distinct stage of object destrudo emerges when at six months the infant begins to develop a prehensile grasp and can hold objects between its thumb and forefinger. This ability has far-reaching consequences. This is the same time at which the infant's eyes can focus on and clearly perceive objects within its environment (Ilg & Ames, 1955; Piaget, 1954). The fact that objects can now be seen clearly and taken hold of gives the infant a sense of mastery and control over objects. Thus, when destrudo is aroused against an object at this stage, the infant seeks to master and control it by taking hold of it from two opposing directions, then tightening its grip on the object. The aim here is to get the object in a vise-like grip from two opposing directions and thus control it. The infant may also put the object in its mouth and grip it with its gums as its mouth closes on it under the power of the jaw muscles or with teeth that may have begun to emerge. The jaw muscles or thumb and forefinger muscles are the executors of pincer-object destrudo.

This new destrudo-energy phase typifies a significant change in the way the infant is capable of connecting with objects in its environment.The infant has embraced the world of objects, outside of and different from its former focus on the self. It is Freud's (1911/1958b) reality principle.

Transformation of the pincer-object destrudo aim is seen in the adult character trait of striving persistently to manipulate and control others. It's also seen in individuals who direct these energies against themselves in reaction formations by constantly, unwittingly putting themselves in dilemmas with others, a pincer situation in which the self becomes the object aim. The opposing interests of other persons are perceived as immutable. These conflicted individuals perceive themselves as being in a vise-like grip between two forces, having no volitional position themselves. These individuals put themselves in such situations unconsciously. For instance, if they carry out some action, a friend or relative will be happy, but another friend or relative will not. They are damned if they do and damned if they don't.

The transition from muscular destrudo expression to verbal destrudo must be noted. Once object destrudo clearly comes into existence, verbal expression toward its aim also occurs. In this sense the vocal apparatus has parallel emotional expression with the muscular expression and development of destrudo starting from the time the infant began to scream and cry when angered during the first six months. This muscular developmental expression is primary for destrudo and the verbal expression is secondary in its expression of muscular intended activity.

Toppler Object Destrudo
(One and a Half to Two and a Half Years)

During this period the child extends its mastery of objects in its environment from gripping it in a prehensile grasp or between its teeth to moving the object to desired places. The child's muscular capabilities lets it move objects around in the environment. Children can now realize the possibility of moving an object to where they want it in relation to themselves. "He may treat other people, especially other children, more as if they were objects than people" (Ilg & Ames, 1955, p. 33). The child's developing musculature can push or pull an object to a desired location (Gesell et al., 1940).

The toddler pulls drawers out or pushes them closed, builds a tower of blocks, then knocks it down. Anna Freud (1972) noted this curious paradox. The child appears to find pleasure in both building the tower and knocking it down. Its construdo builds the tower; its destrudo knocks it down. For the first time in its life the child is beginning to sense the existence and influence of the two drives within itself and their two opposing energies. The knowledge is pleasurable because it provides a true sensing of both drives within the self. Some objects can be pushed and pulled over at this stage; these actions are not only random muscular expressions but a way for the child to gain a feeling of mastery over objects, by changing the relationship between itself and objects. The child's musculature can now enable the child to achieve these ends.

Gesell et al. (1940) noted the pleasure a child finds in mastering relationships between itself and objects in its environment. "For some time he has been able to push a chair around; now he can pull a wheeled toy as he walks" (Gesell et al., 1940, p. 30). The destrudo aim at this stage differs from the first object-destrudo aim of wanting to gain control of an object by getting it in one's hands or grasp. This stage is in the toppler object destrudo stage, the aim here being to change the relationship between the self and object at will by pushing and pulling the object to desired positions in relation to oneself.

Transformation of the muscular destrudo unconscious aims of the toppler object destrudo stage can be seen in adults who develop the character trait of upsetting others or putting them off balance, then moving the person to a desired position. It is pulling the rug out from under the other person, and then allowing the other person to find a footing where the destrudo-instigating individual controls the footing. For example, "You say you are against murder. Then how can you support murder by the state? That's what capital punishment is." If an individual grants that capital punishment is murder by the state, he or she will be on very shaky ground. The destrudo-instigating individual will win the argument using the characteristic trait of toppling others in confrontations. This trait is seen in individuals who are described as being "outspoken," because what they say is designed to undermine their adversary's position.

When the unconscious destrudo aim of this stage becomes self-directed and is transformed into adult behavior, we may observe individuals who always undermine themselves. They quickly admit their faults, thus inviting someone whose object destrudo is in their direction to attack them. For example, these individuals may be fiercely competitive, yet they may show a weakness to their competitors that leaves them vulnerable. Unconsciously, they may work against themselves in competition. They are business people who fiercely compete with their competition. Then through oversight or a blind spot in their business, they allow their competitor to overturn them, or topple them from their position in the competition. It's as though they did not defend themselves against attack that could be forced or waged on them. This is self-toppler-destrudo.

Striker Object Destrudo
(Two and a Half to Five Years)

Now children can verbally express hurts, frustrations, and resentments caused by others. They are aware of their lack of mastery of objects or others in various hurtful, frustrating, and resented situations. Parents and significant adults may discourage them from expressing their resentments. Changing the position of the object in relation to selves does not solve the problem as it did in the toppler years. Now they want to hurt an object or person as they have been hurt by it. They want to frustrate an object as they have been frustrated. They want to cause resentment in an object as they have felt resentment. Striking the object to hurt it solves their problem.

Children discover that their musculature has developed the capacity to hit with hands and arms, to kick with feet and legs, and to hurl objects as an extension of their hands and arms (Ilg & Ames, 1955). The child's muscular capacities allow it to do things it very much wants to do to objects or people. The child can injure and frustrate adults or other children when its destrudo is aroused. Handedness has developed in the musculature, increasing the effectiveness of the striking hand and arm to hit an object with hand or fist, a stick or hammer, or to throw stones at the object.

The destrudo aim at this stage is to deliver a sudden, strong, and hurtful blow against an object that eludes the child's or person's defenses, causing such hurt, pain, and unpleasure that the target itself elects to change its position within the situation. The child is forcing its adversaries to do what it wants them to do. This phase is a step beyond the toppler aim of being able to change an object's position in a situation by the object's repositioning itself. Destrudo's aim now is to cause the object (or the person) to want to change its position in relationship to oneself, because it is too painful and hurtful to not do so. This is the third aim of mastering an object in situations.

This is the striker object destrudo stage aim. Often, a prelude to striker object destrudo is the activity of poking. This is a way for a person to locate and test out vulnerable spots that can be struck without defense or retaliation. One aspect of this destrudo aim is seen in the character trait of suddenly verbalizing hurtful remarks at others when they least expect it and have prepared no defenses. This character trait is sometimes described as being "brutally frank"; it achieves the aim of striker object destrudo without being perceived as such. It is one way for the transformation of the unconscious aim to gain access to consciousness without repression (Freud, 1915/1957a). An opposite transformation of this striker aim is seen when individuals strike at themselves, as when they say, "How stupid of me," or "I'm just no good."

When a child uses a hammer to hit an object and breaks it apart, that dovetails with the next destrudo stage: The child may see the potential in what occurs when an object is broken to pieces. This is an overlapping of stages of drive aims, smashing the object or self into pieces. The result is: the target is no more, in fantasy or reality.

Smasher Object Destrudo
(Five Years to Six and a Half Years)

The fourth and final stage of object destrudo occurs when maturation of the musculature has advanced so far that it can position an object to be destroyed by smashing it into pieces. This is much more than merely controlling an object, the aim of pincer destrudo. This is much more than attaining a new relationship with the object, the aim of toppler destrudo. This is much more than merely hurting the target and causing it displeasure so the target wants to modify its position in the relationship to the protagonist, the aim of striker destrudo.

By this point the musculature has reached a level of development where the child can jump up and down on an object until the object breaks apart. The child can step on a bug and kill it by smashing it. The child can pull a fly into pieces by pulling it apart and killing it. The child can throw a toy against a wall or on the floor and break it into pieces. The child can rip a book into shreds. The child can pound an object with a hammer until it is smashed to bits (Gesell, Ilg, Ames, & Bullis, 1946). Attacks on an object can now be planned and executed. After such powerful attacks the object no longer exists. The child is now capable of destroying objects in a variety of ways.

The aim of destrudo in its final stage is to break an object into pieces, so it can never exist again in reality. The child can end any relationship between itself and an object forever by mentally annihilating it (e.g., "I'll never speak to her again"). Although Freud (1920/1955a) saw this as destrudo's ultimate aim, and Anna Freud (1949) reiterated it, neither of them presented this as a final stage in a developmental sequence of destructive drive stages, as TDT has posited. The final phase of destrudo energy is the smasher object destrudo stage.

Transformations of this drive aim are seen in adult behavior when a person breaks into pieces any vestiges of relationship with another person. The transformations of the smasher object destrudo musculature aims may involve verbal abuse, as when the attacker calls the target all sorts of names and accuses the other person of a vast range of faults and shortcomings. After such invectives, the relationship may be beyond repair and can never be restored. These individuals hopelessly destroy established relationships. They may also demonstrate self-directed smasher object destrudo by destroying relationships they really want to maintain. After they express their smasher feelings, they discover the other person wants nothing to do with them ever again. Some psychotic individuals jump from high buildings or bridges to smash their own bodies on the ground or water below. They might even jump in front of an oncoming train so that their bodies are torn apart when they are hit.

DESTRUDO IN THE OEDIPAL CONFLICT

At this smasher time a major conflict arises between the child's destrudo aims and its construdo and libido aims. For the first time a conflict exists among the three drives (destrudo, construdo, libido): Oedipal conflicts are at the peak for libido (i.e., sexual desire for or desire for total possession of the parent of the opposite sex), whereas object destrudo wants to annihilate the same-sex rival parent. This conflict causes maddening unconscious conflict with the construdo aims to be combined, connected, and to construct a loving relationship with this rival parent. If the parent blocking the child, the rival parent, is the object of the child's unconscious destrudo annihilation fantasies, then intense conflicts with the child's construdo strivings to bond with the parents are created. Unconscious guilt feelings may be produced in the child's unconscious construdo drive. This unconscious construdo attachment arouses unconscious self-destrudo in the child's sense of self-connectedness.

If the child actually achieved its wishes, then the parent of the same sex, still loved by the child, would no longer exist. Wanting to end the existence of this parent causes terrible unconscious conflict between construdo and destrudo impulses causing intense guilt in the child. Hence, the child's aroused destrudo solution places it in a maddening position that solves none of these unconscious drive conflicts. Unresolved drive conflicts leave individuals at the neurotic level of drive disturbance or the neurotic level as mental difficulty.

CHAPTER 7

AROUSING DESTRUDO OR CONSTRUDO: FACTORS AFFECTING FREQUENCY AND INTENSITY OF AROUSING DESTRUDO

THE AROUSAL OF OBJECT-DESTRUDO

The arousal of object-destrudo occurs when one's construdo fantasies are not met in reality and one experiences blocked construdo expectations. Construdo drive frustrations arouse the destrudo drive. This frustration may be resolved by self-denial in what individuals will allow and not allow themselves. It is seen in the morals, ethics, and standards they create for themselves in the human conscience. Freud (1923/1961c) called this mental principle the superego. In *Civilization and Its Discontents*, Freud (1930/1961a) indicated that destrudo is aroused when the instincts are blocked.

When one or more of the stage aims of construdo is/are involved in frustrating circumstances, then one or more of the destrudo stage aims will be aroused. If a family expects to be treated courteously by their neighbors, but finds they are treated with insults, then striker-object destrudo aims will be aroused toward the neighbors. If children control their parents during adolescence, paradoxically they will grow up feeling they have not been parented. They feel robbed of meaningful, adult-controlled direction and nurturing connections with the parents. These feelings arouse unconscious destrudo against their parents. Their construdo positive parental relationship is blocked by the unconscious destrudo feelings they hold toward their parents. The result is a troubled relationship between the adolescent and parents.

Often, people envision a self-connected construdo plan or fantasy about what they want their lives to be like in the future. They dream of creating a happy marriage, constructing a successful business or engineering career, or becoming an entertainment or sport star. They dream of constructing a means of amassing large amounts of money. When their planned time comes, but their dreams have not come true, for many object-destrudo is aroused. They rationalize their shortcomings in believing that circumstances and people have blocked the attainment of their dreams. They respond with toppler-destrudo toward such circumstances and striker-destrudo toward the people involved. Henceforth, they may show a bitterness and pessimism in their attitude toward life.

THE PATHWAY FROM OBJECT-DESTRUDO AROUSAL TO SELF-DESTRUDO AROUSAL

A change from the arousal of object-destrudo to self-destrudo occurs when individuals are too frightened to express their anger at certain people. They're intimidated by their thoughts about what these others might do to them in retaliation for their anger toward these others. Angered by her father ignoring her, for example, a young girl wants to yell at him for his behavior. She fears, however, that he might stop speaking to her, as she has seen him do with his friends. Such squashed anger can be turned from important others to the self. In such circumstances, individuals begin belittling themselves, because they were unable to courageously express their anger directly at these significant figures in their lives. It can cause them to have feelings of unworthiness, which further contributes to their readiness to attack themselves when angry. It's one type of retreat from the conflict between wanting to show anger and fearing the consequences.

At a neurotic level, people may repress the anger; that is, block it out of all conscious awareness. It is pushed mentally into the unconscious. The anger might reappear or be displaced from the unconscious to be expressed at someone with whom it is safe to be angry. It could also be expressed by taking an antagonistic political or ethical position against a group.

At a more troubled level, patients with borderline personality disorder in a thorny confrontation with another adult might regress to an earlier emotional level. Borderline patients often regress to the thrasher destrudo stage, in which individuals attack themselves when attacking an object outside of them. This is because at the transitional object/self stage, both the self and the other were targets.

CAUSES OF SELF-DESTRUDO

Self-Destrudo Reactions to Being Defeated

Some individuals react to frustrated and ungratified drive aims with a self-destrudo reaction. When individuals' construdo anticipates losing things that are very important to them, they may attack themselves for the anticipated construdo loss. Energies of toppler self-destrudo are turned against the self. When a person's construdo anticipates being defeated in a confrontation with another, it arouses object-striker destrudo at the self for being vulnerable to the anticipated poking and striking from the other. It is seen when individuals prefer to kill themselves rather than wait to be killed by their enemies. An escaped slave would rather kill her children than have them returned to slavery when she's caught by her owner. Object-striker destrudo is the only drive energy she can employ in this situation to express her will.

Thus, destrudo aims are the woman's only drive outlet. Although she's limited by the situation, she can still assert herself and express her construdo defiant wish by her destrudo drive choice. This is also seen when a person would rather commit suicide than be taken by the law. Similarly, troops in battle at times would rather fight to their death than be taken as prisoners by the enemy.

EARLY FAMILY INFLUENCES
ON THE AROUSAL OF SELF-DESTRUDO

During individuals' early family relationships with their mothers and fathers, with relatives, and with siblings, and in an adolescent's social relationships with peers and later with adult friends, associates, and colleagues, some individuals may experience contradictory, opposing viewpoints to their union with these people. Such influences of others may oppose one's positive construdo connections. Instead, these individuals may deepen regression following the initially aroused destrudo.

From the First Trimester

The embryo's first biological, biochemical, and biophysical reactions to the possibility of its death occur during its first trimester in the uterus. Some failure of the mother's body to maintain a proper hormonal balance, a defective genetic factor, a disease process, or an external injury to the embryo can cause a spontaneous abortion or miscarriage. If this process abates before miscarriage, the embryo returns to normal func-tioning. Nonetheless, therein begins the human organism's sensing of self-destrudo – factors within itself that can undo its construdo combining cells, cell groups, organs, and organ systems for maintaining and continuing life. Herein begins the embryo's resistance to self-destructive processes within itself. When these factors are in time outlived, the embryo survives. Outliving such destrudo influences against the self can therefore become a survival solution in dealing with such influences in oneself.

The fetus registers the mother's emotions only as a sensation of feelings and reactions. Thus, these emotions are not remembered as such, but only as sensing/reacting patterns toward the self from an unknown source. In its entirety, the break-apart self-destrudo aim is the most devastating and most basic self-destrudo aim. It is what is returned to by the principle of returning to undermost energy origins. The fetus may simply sense a need to totally destroy itself.

By the end of the second trimester, all of the fetus' organ systems are fully formed, including its neural system. At the most rudimentary or earliest form the neural system of the fetus can sense the pregnant mother's response to its development within her. Even at this early space-time, the fetus can sense the mother's response to its existence: wanting, welcoming it. She might experience the fetus' presence in her as a pleasure or joy. She might reject her fetus as unwanted and of no value to her or to her life.

Second Trimester Fetal Self-Destrudo
Caused by the Mother's Rejection

During the first and second trimesters the pregnant mother may have many reasons for rejecting the infant developing within her. By the end of the second trimester and certainly in the third, her feelings about her pregnancy are defined as disorders. These individuals often become aggressive in asserting their value and worth. They become very competitive in trying to stay ahead of others in proving their value. They do so through object toppler destrudo of others.

It is proposed by Triadic Drive Theory that a psychopath's self-construdo starts in utero to combine with the mother's uncertainty about the fetus' value to her, then disconnects from these potentially rejecting sensations, and thereafter combines with its own sensing of itself and more pleasant sensations from a more positive combination. Following birth, the newborn experiences its value in returning to this original relationship with itself. In psychopathic individuals, self construdo evaluations only identify and incorporate their own feelings about themselves, not those of their parents as in mentally healthy people. However, psychopaths do seek reinforcement (from others) of their own self-construdo's positive evaluations of themselves. These individuals are overly or excessively self-connected.

The self-construdo of a person with borderline personality disorder cannot distinguish between its registered hostility to itself that presumably stems from its mother's hostility toward it and its own aroused self-destrudo toward itself. It cannot determine with which to combine because its self-construdo lags or is as yet undifferentiated in its unconscious form of object construdo. The borderline patient experiences a confusion that is never resolved. In psychopaths, self-construdo can be differentiated from self-destrudo, whereas in borderline individuals self-construdo cannot be differentiated from self-destrudo.

Inexplicable depressive reactions, sometimes with suicidal aspects, trace directly to times when a pregnant mother felt she did not want her child. In adulthood depressives can replicate the mother's feelings with similar feelings toward themselves. Exactly where in space-time construdo suddenly links ongoing circumstances to these uterine rejections is the inexplicable or uncertain factor in such a person's depressive reactions.

The frequency and intensity aroused in these three conditions of construdo and destrudo (psychopaths, borderlines, and depressives) all relate to the strength and extent of a conflict in utero. Consequently, the uncertainty principle applies here in terms of predicting future development of any of these psychopathologies.

The Third Trimester and Fetal Self-Destrudo

In the third trimester the fetus may experience the self-destrudo aim of a defensive part causing self- destrudo. By the third trimester all three drives are differentiated in the unconscious. During this trimester, a genetic defect in some part of the fetal organism (heart, kidney, liver) can cause the fetus to die or not survive for long following birth. It is a prototype for psychotic depressive and manic-depressive episodes when individuals blame themselves unreasonably (over-connect) for what they failed to do (exaggerated guilt) or should have done (exaggerated standards or superego self-punishment), evoking self-destrudo pincer, toppler, striker, or smasher – for not being perfect.

The pregnant woman's rejection of her future newborn may be caused by her fear or dread of the birth process itself. Stories she has heard or read about dying during childbirth can contribute to the pregnant woman's bad feelings about her condition. These are mainly construdo bad fantasies but occupy a reality corner of her mind. It is negative construdo anticipation or self-destrudo anticipation. It becomes a real and constant anxiety.

A delivering mother may direct outright hatred at her child as the result of a psychosis – the mother's hatred of her own life that she cannot bear to see passed on to her child at the space-time of the delivery. It is a postpartum psychotic reaction. In TDT, the newborn unconsciously can incorporate a sensing of the mother's hatred. It is a model for self-hatred later in life. This self-destrudo may be activated when one feels intense anger from a loved one. The interpersonal situation resembles the partly loving mother's feelings toward the fetus, with hate present as well.

Self-Destrudo Caused by Birth,
Arousing Traditional Construdo and Destrudo,
and the Original Anxiety

Birth shockingly and painfully removes the newborn from the warm, fluid, lightless, silent, self-contacting, nutriment-giving, waste products-removing environment to which it was self-combined and connected. The newborn is thrust into a new, cold, air-filled, light-filled, sound-filled contactless environment, immediately arousing the newborn's destrudo because it is construdo unexpected. It is the unfulfilled construdo expectation of a continuing environment that suddenly ends and thus arouses destrudo.

However, this aroused destrudo has no immediate object aim at this point in space-time except the self-destrudo aims from uterine times. Consequently, the aroused destrudo is mostly directed at the self by the defective part causing self-destrudo aim. Nevertheless, in the abrupt transition from the uterus into the world, some construdo automatically begins combining with the new environment in this transitional construdo stage. This combining pulls some destrudo with it in opposition to the environment in the form of the transitional destrudo aim.

Birth, this initial major transition in the human organism's existence, produces conflict in the unconscious and drives uncertainty within destrudo as to what should be its aim – the self or the new environment. Conflict and uncertainty are also produced by construdo, in terms of how to combine with the new surroundings and how to disconnect from its uterine world. With respect to TDT (Triadic Drive Theory), the newborn is automatically influenced to do so by the underlying binding forces of physics creating construdo.

This alternation in construdo attachment is a major event in the construdo experience. It's the newborn's first encounter with changing construdo ties. It's the first sensing that it can disconnect from one's life experience and join another. This produces changes in the newborn unconscious that radically disrupt the old uterine balances of the two drives' energies (construdo and destrudo) as they begin to differentiate or separate during the end of the third trimester into two focuses of the unconscious. First, the self stages of construdo and destrudo; second, the object environmental directions of the two drives. Birth is a radical change from previously balanced destrudo and construdo unconscious energies in utero, versus the separating drive differentiating after birth in the second focus of the unconscious.

At birth, self-construdo has developed through four uterine stages culminating in the self-creative construdo stage. Destrudo has developed through two uterine stages culminating in defective part causing self-destrudo. The third drive, libido has developed through two uterine stages culminating in the contactual stage. The balanced drive system, the interrelated energy forces achieved in utero by these three drives, is completely disrupted at birth.

The transition to life outside the uterus forces the newborn's emerging unconscious and its differentiations into three distinct drives into a state of disarray and uncertainty regarding the drive aims (space), and time directions (backward or forward in space-time). This does not happen again until the Oedipal conflict, when all three drives are put into a state of intense interactive conflict and disarray. Evidently this process of transition from uterine life to birth creates a mixture of conscious and unconscious drive conflict.

Whether the drives seek object aims (space) in the new environment or from the old environment of uterine-dependent existence, or whether the drives will be directed forward or backward in time – is comprehensible by the uncertainty principle. Construdo determines the forward time direction of the three drives. Destrudo determines the backward time direction of the three drives. Libido operates between the two. Libido has previously been satisfied in its contactual aim during the third trimester, which continues following birth. Holding, touching, stroking the newborn and covering it with a blanket, satisfies this aim in the newborn during the transitional phase after birth until oral libido begins at about six months of age, when much of the surroundings are understood by the newborn who puts everything into its mouth.

The birth transition arouses uncertainty for the drives. A mood of anxiety is first sensed by the just born. More drive discharges occur than the newborn can assimilate. The newborn is overwhelmed by its three drive firings. For triadic theory, there's a major transitional anxiety found in the reluctance of construdoto join into these new surroundings. This causes initial uncertainty for all three drives.

Anxiety, remember, is the fantasized destrudo event gaining predominance in the individual's imagined future – construdo future! It is expecting bad, destructive, things to happen – destrudo things!

During the transitional drive time, libido follows first aroused construdo and then destrudo. Before birth satisfaction of libido lay in the contactual aim. After birth libido is radically changed from its uterine satisfactions. Now touching and bodily contact of caring adults continues the contacted satisfactions in a new way.

Another cluster of libido joined with construdo is sense by the newborn with respect to the cold air around it, the disappearance of warm fluid, the change to a lighted, sound-filled environment. Nutriment serving and waste eliminating are now a function of its body, all of which are part of its new world. Libido has a destrudo-undifferentiated response to the new environment, which is reinforced as the newborn experiences the new world. Gradually, the contactual aim becomes secondary to libido's nutriment-biological needs and the mouth becomes a focus of psychological drive satisfaction in the oral aim of libido derived from its receiving nutriment.

All these changes challenge the newborn's differentiating drive responses from the unconscious for adaptation and survival in the new world. The conscious experience of this disruption of the three drives continues until they restabilize or rebalance, or else results in construdo's reluctance to combine and connect with the surroundings, which causes anxiety. This transitional stage anxiety indicates that the drives are undergoing transitions when, in the newborn, they enter the new world.

While Freud (1926/1959b; 1933/1964) saw birth as generating the first anxiety, he disagreed with Rank's (1929) concept of birth trauma and the intensity of the influence on neurosis throughout an individual's life. TDT endorses Rank's description of the overwhelming drive conflicts, as birth is experienced as trauma. TDT considers birth the first basic anxiety for humans in the new world. Freud considered childhood anxieties such as fear of strangers, of loneliness or being left alone, or of leaving home (as in separation anxieties), as threatened libidinal attachments to the mother. Anxiety from these situations caused repression (Freud, 1933/1964). But all of these situations clearly involve threatened construdo attachments – the drive that was not in Freud's thinking. He considered castration anxieties in males and feelings of having been castrated in females to be derived from partially released repressed libido. His conclusion did not consider the destrudo aspects and the fantasized implications of construdo in the newborn's situations.

Essentially, transitional stage anxiety occurs in new or changing reality situations when there is opposition or conflict of direction and object aim within one or more aroused drives in the unconscious, causing varying degrees of uncertainty from the drive energies that reach consciousness. These drive energies may be displaced, condensed, reversed or symbolized when they enter consciousness from the unconscious. The varying degrees of uncertainty depend on the number of drives involved and what they have been aroused to do, whether they react in a forward or backward time direction or take an object or the self as their aim. These are the drives of construdo, destrudo, and libido.

TDT considers the extent of the three drives' involvement as paramount in the production of anxiety. It is reluctance by the drives to move forward and fully connect with the situation. An individual experiences transitional anxiety whenever the conflict to join or not to join is intense. This is because of disruptive changes in the drives' otherwise natural pathways in the designated situation.

Freud (1933/1964) saw birth as "our model for an anxiety state" (p.93), but for very different reasons. He considered "signal anxiety" as a fear that a model might be repeated or re-experienced by the repetition compulsion as the second origin of anxiety, quite different from TDT's notion of construdo drive reluctance. Later, we will see that drive interactions and frictions between the drives' energies in interactions cause moods and emotions.

Self-Destrudo Arousal
When the Mother Has a Threatening Delivery

When a delivery is prolonged and difficult and the mother's life is in danger, there will be moments in the mother's unconscious when her wish is to rid herself of the infant, because its survival is threatening her life. As a person her unconscious wishes it dead. If the child dies she will live. As a mother, however, she does not want the unconscious destrudo wish aimed at the infant to be realized.

Hence, both her construdo connection with herself as a mother and her unconscious destrudo are locked in an intense conflict. In Triadic Drive Theory the life - death conflict of the mother is transmitted to the newborn. It thus unconsciously senses how it can feel toward itself later in life. It carries in its unconscious the mother's death wish toward itself. This is one origin of self-destrudo, self-hatred in later childhood, adolescence, and adulthood. Such unconscious self-hatred can be touched off when individuals experience hatred from persons on whom they are most dependent and love. That's because theoretically there is an emotional similarity to the emotional relationship with the mother in utero.

The newborn can incorporate its sensing of the delivering mother's unconscious smasher- and striker-destrudo aims toward itself. It comes to introject and identify with the hostility of the delivering mother. Such intense self-destrudo expression, when later released, can cause despair or a depressive mood in the adult. Occasionally, fantasies of suicide may appear. In accord with the delivering mother's destructive fantasies, these suicidal fantasies take the form of self-smasher destrudo (e.g., jumping off a bridge) or self-striker-destrudo (e.g., shooting oneself in the head). One cannot predict which of the above potential possibilities might occur. The outcomes are subject to the uncertainty principle.

In a threatening delivery the infant seeks to continue combining with life, while the mother unconsciously wishes to end its life. These two influences are opposites, coming from two different directions in regard to the infant in the birth process. This conflict of mother versus newborn can implant in the infant a sense of being in a bind or trapped between its conflicting wishes. This is the pincer self-destrudo aim, and the self is the target. The pincer self-destrudo aim is what the infant carries into later life and may employ when destrudo is aroused. One such pincer self-destrudo attack can be seen in addiction. An addict is trapped between the need for a substance that's destructive versus a construdo wish to combine with it more and more.

Self-Destrudo Introjected From Parents'
Destrudo toward the Child

When caring adults express mixed destrudo aims toward their infant, the infant will often incorporate these aims toward itself during the first six months of life. That is in the drives' transitional phases, when construdo is combining with whatever is in its environment. External reality destrudo against the newborn may take the form of smasher-object destrudo; a psychotic parent may express smasher-object destrudo against an infant by picking it up and throwing it down in its crib as if to smash the infant into pieces. Most likely, this adult had been similarly treated in its infancy and learned or construdo-connected with this method of dealing with a fretful, annoying, and provocative infant. Striker destrudo against the infant can be seen when an angry parent slaps the infant child, despite its tender age.

Toppler object-destrudo against the infant's expressions of displeasure can be seen when the caring adults' attempt to pit the infant's expression of dislike and discomfort against the infant. This happens when the infant cries too much and too long, to have its diapers changed, to be covered when it is cold, fed when hungry, or put to sleep when tired. The adults refuse to do what the infant wants if the infant is too demanding in its efforts to achieve its ends. This infant learns to temper its needs and its expression of these needs to what the adults around it will respond to. The concept of a reciprocal relationship is being imposed on the infant, versus the infant viewing itself as the only concern of these adults. The example above can also be seen as pincer self-destrudo if the infant senses that it has a need to cry, and that its crying is causing the trouble it is having with the adults. It is thus trapped by its own response, which is more than just adjusting to what the adults want.

Concluding Thoughts

The frequency and intensity of the above self-destrudo expressions experienced in later life will depend on the extent to which infants were frequent early victims of adult attacks. Incorporations and identifications of the adults' attacks on such infants occur, and the infants may later, as adults, intermittently feel hate for themselves or otherwise undermine their own efforts to promote themselves. This destrudo is connected to the self by construdo in an earlier space-time. It can be reactivated in later space-time when emotional currents or aspects of the present reality circumstances parallel or are similar to the earlier conditions. This parallel of earlier experience to current experience causes a transference reaction, so called because the present-day response is a reaction to stimulus conditions as if they were the stimuli conditions of old. The reaction is transferred from one space-time to another.

Remember, the unconscious is timeless. It does not consider whether an arousing stimulus occurred six months or sixty years earlier. The response transferred from one time to another will be the same.

AROUSAL OF OBJECT CONSTRUDO

Diverse, discrepant stimuli in objects, events, people, and personal experiences that can be put together or combined into mental categories, with common links and fantasies of future psychic intentions are the stimulations for construdo arousal. Construdo energy pulls and connects these categories together into units that get bigger and more complex. The infant's perceptual sensory system is primarily involved; that is, their auditory, visual, olfactory, gustatory, heat, cold, touch, and kinesthetic sensitivity.

Similarity of sounds, tastes, smells, or physical movements, as well as proximity of these factors, are important in the groupings or categories of these stimuli. A negative, unpleasurable sensing or feeling can also be a bonding or combining influence that can set the tone for similar future negative object relationships, or similar circumstances in which these object relationships occurred. These sensory impressions are the beginning of the infant's sensory, memory-like traces. These sensory images have no symbolic language memory, but they are retained in the infants' unconscious and influence them throughout later life from they know not where.

AROUSAL OF OBJECT-COMBINING CONSTRUDO

Stimuli, objects, events, and personal experiences occurring close together in space-time along lines of sense modalities arouse construdo to group them together at the earliest times as construdo combinations. The aim to combine with the world works forward in time. It can also work backward, but that requires thinking the infant does not yet have. When objects or experiences, such as touch and kinesthetic stimulation in the way the parents hold the child, show a similarity to each other, then construdo energies will combine them.

As a human organism develops, its past experiences, preverbal sensory experiences, and later memory traces all begin to affect construdo's tendency to combine and connect things. This is a notable moment, for it indicates that memory traces begin to be affected by construdo operations. Construdo will be aroused to combine with whatever is available. This makes construdo combinations out of salient, distinct, and clear psychic impressions, registering them as seen in the beginning of this section, which follows the grouping principles of Gestalt psychology (Koffka, 1935).

Object-Connecting Construdo

The father's visual image, his soft vocalizations and his laugh, the kinesthetic pleasure or unpleasure as he tosses the infant over his head or swings it between his legs (while being held and stroked in contactual libido by the father's warm arms and hands), all arouse the infant's construdo to connect these diverse sensory impressions of the father with itself. The parent may provide all of these sensory impressions. The infant's "dada" is its first vocalization of a new symbolic language, connecting these reality sensory impressions with the reality presence of the father. The infant attempts to connect with all that has been registered in the infant's psyche about this figure, as construdo connection aims are aroused by the father's stimulations. Construdo object-connecting stage aims continue up to the age of one and a half years.

Object-Constructing Construdo

After a year and a half, sensory impressions are construdo com-bined and connected with verbal symbols or words and perceptions of stimuli and objects in the surrounding environment. The child now perceives and is driven by construdo arousal to put together these sensory impressions, verbal symbols, and object perceptions: The child sees what can be constructed by putting them together. Past experienced sensory impressions and perceptions have produced memory traces that now arose construdo to put them together with present or ongoing events. In so doing a new construction is produced.

The child's activity and actions in reality arouse its imaginative scope for being combined and connected with what it will do in the future, giving rise to construdo fantasies, plans, or constructions of what the child will do next. These constructing aims will continue from one and a half to three and a half years.

Later in life, the frequency and intensity of such constructing aims will depend on whether the child experienced success and approval during its formative years or the frustration and discouragement of being unable to complete such aims. Parental, sibling, and peer indulgence, approval or disapproval, and discouragement, can lead to the later child and adult construdo's wanting to do more or less of these constructions.

Object-Creative Construdo

From three and a half to five or six years of age, the child develops creative or entirely original ways to put together impressions, images, perceptions, and memories of past experiences, as well as the impetus to construct new constructions out of all of these previous factors. Beyond this aim, after three and a half years of age, the individual is aroused to do more, to create from its construdo driving energy its own original, unique perceptions of the world. The new creative stage aim stems entirely from the individual's own mind. This creative stage aim aroused when no more can be seen or envisioned as old, established ways of putting together these diverse, discrepant stimuli. Outside of these bondings with reality influences is the imaging of one's own reality. The child sees no further possibility than creating its own reality in imagination or describing and defining it as the child wants to see and imagine it.

The construdo creative stage aim lies behind the grouping Gestalt principle of connecting stimuli toward a "good figure," or *Prägnanz* (Koffka, 1935). An individual connects stimuli by what is the most creative combination of its perceptions of the total situation. These situations involve arousal of construdo and its creative aim – putting things together in the most original, unique, creative way.

Opposing construdo's self aims and their arousal are underlying self-destrudo aims, breaking apart or attempting to destroy the individual's self-connections, self-constructions, and self-creations, when we attempt these things in ourselves. Such self-destrudo resists self-construction, self-creation images of ourselves. Rather, such self-destrudo would tear us down.

Freud (1933/1964) saw this as the destructive drive turned against ourselves, causing resistance in psychoanalysis to understanding ourselves. The frequency and intensity of such expression depends on the earlier frustration of the self-construdo aim or its overindulgence, either one making these aims hard to relinquish.

In Triadic Drive Theory (TDT), deep resistance to ourselves can be a reflection of unconscious self-destrudo. As such, self-destrudo can weaken our immune system, making us vulnerable to contracting different illnesses. Again, according to TDT this unconscious self-destrudo can lower our resistance to a cold or the flu. When we think about (yet dread) a given disease or condition, the construdo connection of these thoughts builds a pathway that – if traveled by our nerve impulses too often to the actual physical site – can lead to our contracting the disease or illness. It is a punishment of ourselves. Fearing a heart attack can make us more vulnerable to having one under times of great stress. It is a striker self-destrudo attack on us.

AROUSAL OF SELF-CONSTRUDO

Self-construdo arousals can be seen when individuals want to join others around them in some way but see that some alteration in their displayed personality will be necessary for success.

Self-Combining and Self-Connecting Construdo

Self-combining and self-connecting construdo may be employed when individuals want to make some direct connection with people they want to befriend or relate to in a positive way. It might be showing an interest in getting into a basketball game with some guys in the playground by asking them how one gets into the game. It might be smiling at a possible partner when asking for a dance. The person is self-connecting with a positive reaction to others and showing it in a smile. It might be giving candy to children in a foreign land to befriend them. We know children like candy and are using this knowledge to facilitate making friends. In these instances we are combining and connecting with certain personality aspects in ourselves and using them to gain positive reactions from others.

Self-Constructing Construdo

The arousal of self-constructing construdo can be observed when peer pressure causes teenagers to construct and conform with an image given them by peers. They might be called "jocks," and begin to act like their only interest lie in sports during their high school years. They construct from their athletic talents and interest an image that conforms with their label "jocks." Self-constructing construdo is seen in the focus on "toughness" and "we can do it" attitude in the image embraced by recruits constructing themselves into Marines. It is seen in the conservative dress and reserved manner in the trainees shaping themselves into "bankers."

Self-Creative Construdo

The arousal of self-creative construdo is often seen between two people in a love affair. They will change themselves and adopt characteristics desired by the other, and they may directly copy characteristics found in the loved one. They may acquire an interest in antiques, movies, or football because they want to share the loved one's interests. They create or mold characteristics in themselves they didn't have until they thought it would deeply please the loved one.

Concluding Note on Self-Construdo

Self-construdo leads to various types of self-building and self-invention. It's an extensive psychic activity that has not been a focus of psychology, but is nevertheless revealed in a variety of human behaviors. TDT has focused attention on this activity, as well as given it a drive basis.

CHAPTER 8

FREUD'S DUAL DRIVE THEORY CONTRASTED WITH TRIADIC DRIVE THEORY

In conceptualizing an instinct or drive, Sigmund Freud (1915/ 1957a) asserted that an instinct affects individuals and leads them toward specific goals of which they are completely unaware. Freud's assumption is that the aim of a drive is unconscious. Destrudo's ultimate aim is to break an object apart, into pieces, so that its existence in the world ends forever. The individual's self can be the object of destrudo, or more commonly, the object is another person or thing. Destrudo's aim is tantamount to reaching a state backwards in time when the object was nonexistent.

For humans and their self-smasher destrudo that ends in death, this means nonexistence on the planet and in the entire universe. This end is tantamount to the conditions at the time of the Big Bang when an energy singularity exploded into matter, creating the universe and destrudo energies; no life yet existed in any form. This is the state of absolute death. Later, hydrogen and helium atoms came into existence and evolved into carbon and oxygen atoms that became a form of the universal organization, changing again into living molecules and organisms. Each of these levels of organization in the universe existed as earlier states of what later became human life on earth after matter in the expanding universe solidified into bodies. We go back to these earlier states or levels of universal organization when we die and decompose again into atoms, a process that takes thousands of years. Nevertheless, the process approaches the principle of returning toward undermost energy origins, or to earlier states of existence before there was life – the ultimate aim in the psyche of destrudo against the self.

These backward progressions (or retrogressions) in states are matters of physics. Changing from one state of existence to another (e.g., ice to water to steam) is a matter of physics. Since Triadic Drive Theory (TDT) is based on physics, it facilitates a different consideration of the physical states found in Freud's death drive.

When Freud speaks of a return to the inorganic as an aim of the drive, TDT translates this as a return to atoms. In the space-time of the universe this means to a time when there was only atoms in the universe, carbon and oxygen, hydrogen and helium. This is the universal space-time of what TDT calls absolute death, before there was any life in the universe. It is a backward progression (or recession) to earlier states from which life was made. The progression is backwards in time to smaller and smaller components. It is the same as the ultimate aim of destrudo. Beyond universal time-space, it can be seen in a single organism's biological lifetime, when destrudo is directed at the self.

When a fetus exists in utero, if its mother has a miscarriage, then the embryo or fetus breaks into pieces that are expelled from the mother's body. This is an example of the unfolding of destrudo's ultimate aim and influence on the state of the organism.

When destrudo is reversed or turned outward against the environment, its ultimate aim is to break apart an object into pieces so it can never exist again in reality. The destrudo object can be a person or thing. The destrudo-driven fantasies may include blowing up the other person, but this fantasy is seldom carried out in reality. When the target object is inanimate, we often see someone hurling an object to the floor or against a wall to smash it to pieces, or jumping up and down on the object to break it apart. These actions end the existence of the thing. In effect, this moves the object back to a time before it existed. These actions thus change stages toward backward in time.

Conversely, the ultimate aim of construdo is to unite individuals with their environment – to combine, connect, and construct a bond with that environment or create a unique way in which these individuals can be connected with their environment. The drive's aim is not conscious but unconscious. People do not consciously understand that they are impelled to join other individuals or things in their environment or the environment itself. Freud never considered such a drive. He attributed such actions to sublimations and derivations of libido and the operation of cathexis.

Living creatures are products of the universe and its forces, which hold everything in it together and cause things to operate as they do – including us. It is proposed that we are driven by energies of physics in large measure, not biology, as Freud believed. The concept of the three drives links us with everything in our universe through binding forces of physics, as well as its destructive forces. We come into the universe originally and leave it by changes in states of existence – a matter of physics, not biology alone, in our various conceivable ultimate endings.

THE THEORY OF THREE DRIVES

Human behavior and psychic ideation is driven, TDT asserts, by libido, destrudo, and construdo. The last two are created by the same forces of nature that created the universe. Without them, humans could have never come into existence.

Biology could not have existed without these forces. The first inorganic atoms would never have held together, nor would they have combined with other atoms and held together to create the first living molecule. Every living thing, humans included, is a product – a creature – of the universe's forces, above and beyond biology. As creations of the universe, our psyches are governed predominately by the same principles as govern the universe. Biology did not create the universe, nor are biological principles those by which the universe operates. Physics exists in the universe without biology, but biology cannot exist without physics. The principles of physics are universal. Those of biology are not. When he detailed the libido drive Freud (1905/1953b) first observed how a drive influenced the human psyche from birth to death. Sometime later, Freud added a second drive based on observations of human beings with respect to the destructive drive (Freud, 1920/1955a).

Drive theory is put in its most meaningful place when its forces are based on and congruent with those of the universe. In introducing construdo we have attempted to show that the forces of physics created a human drive that determines most of the force of the human psyche. Libido is a secondary human drive derived from biology, but the influences of biology had to be observed before greater insight could be obtained into the more fundamental unconscious human drives here proposed by Triadic Drive Theory (TDT) to be derived from physics.

Earlier, Freud (1900/1953a) in his classic work, *The Interpretation of Dreams* described and detailed how the psychic drives stemming from an individual's unconscious or the id might seek access to the conscious mind. He described how a drive might be repressed, blocked out of the conscious mind, yet seek to return into consciousness. Freud explained that the unconscious had a timeless reference in terms of events; reversals of the manifest content of dreams might also reflect beneficial assertion of a dream's message. He described how transformations of dream content might reach consciousness. He delineated displacement of an aim, fusion with the other aims, symbolization of an aim, and so on. He gave us a method for understanding dream messages, in terms of how a drive idea found its way out of the unconscious.

Triadic Drive Theory asserts that the actions of subatomic particles are similar to the principles of the unconscious operations. Chester (1978), a physicist, observed this similarity in his book *Particles.*

Construdo and destrudo are derived from forces of physics that oppose each other. These forces are found in the cells of all living organisms, namely the catabolism and anabolism of cells. Construdo behavior advances an individual forward in the environment in time, while destrudo behavior has the ultimate aim of undoing what has been achieved, regressing situations back to before they began – backward in space-time.

In the id, construdo and destrudo are opposing energies that produce unconscious conflict. The reservoir of construdo energy in the id or unconscious psyche is almost twice as much as destrudo throughout most of a person's life, and about equal to destrudo and libido combined. Consequently, construdo's influence on behavior generally predominates. Although there can be occasional oscillations to destrudo and its expressions.

As the construdo energies come into consciousness – into the ego – they accumulate and function as conscious drive energy from the ego. This has been called the synthetic ability of the ego to promote joining the environment or putting things together (Hartmann, 1964). "Putting things together" is the drive aim of construdo. However, as construdo energies are released into the ego, there is a corresponding pull from destrudo in the id to separate from them.

When people are primarily hostile to certain groups, they will reveal breaks in their hostility and think of somehow connecting with their enemies. When people are supportive and nurturing toward certain groups, they may experience momentarily hostile thoughts and fantasies toward the same group. Some individuals are tender one moment and hostile the next. From one generation to the succeeding generation, construdo and destrudo may be seen to alternate. A minister's son becomes an outlaw. What is advanced to positively connect with society may be pulled down subsequently, with or by others.

These alternations between the two drives can be seen as drive energy going forward in time versus drive energy going backward in time in reaction to different events. This drive reversal in energy through time is similar to the way subatomic particles can change (Chester, 1978). How these particles change direction or the nature of their energy and identity underlies in the unconscious the unpredictable changes from construdo to destrudo. Destrudo may return to construdo then possibly change back to destrudo. Psychotics, who have little rational conscious psyche control, are likely to demonstrate this unpredictability in their expressions of construdo and destrudo. Physicists explain that subatomic particles also have anti-particles or particles that can annihilate each other when they are combined (Hawking, 1988). These subatomic energies operate in opposition to each other, just as construdo and destrudo exist and operate in the unconscious.

The hold construdo energies have on destrudo energies, by their combining influence in the id, can produce transitory repression or retardation of certain quantities of either drive energy. The repressed amount of the drive is held back from consciousness in the id by uncon-scious construdo – it is stronger than destrudo.

Parental disapproval of destrudo expressions or teachers and parents belittling or demeaning construdo efforts and expressions can produce displeasure and the repression of either drive's energy expression. Parents can feel threatened by a child's energies to join and connect with the world in ways outside the parents' influence. A child's construdo efforts to connect with the world may show no resemblance to the way the parents' construdo energies have joined the world. Consequently, these parents oppose, stifle, or attack their child's construdo efforts, causing the child displeasure. If the child then incorporates an opposition to its own construdo in the superego, it has taken into itself an opposing feeling and motive toward its own construdo. These young people become adolescents and later adults who operate within their own psyches in opposition to themselves. Whenever they sense their construdo-drive energies motivating them to connect with their world in the way they had planned for themselves, they also experience the parental influences that opposed and belittled these efforts. Therefore, they hold themselves back in the world, without attempting to reach their true potential.

When this situation exists between parents and adolescents, the adolescent's pushed back construdo may cause an advancing destrudo. These adolescents may become hostile and antagonistic toward an adult authority. Under its own energies, destrudo may then become active in its expression against parents until the parents can no longer influence the adolescent because the adolescent has experienced their presence as counterproductive to its own construdo. This oscillation or alternation between defeated construdo and ascendant destrudo can lead to self-attacking destrudo.

Because Freud's libido drive is directed by biological forces, it is outside the conflict in the id between the two drives directed by the forces of physics. To Freud's (1905/1953b) libido stages after birthwe added before birth the forming libido stage during the first six months in utero and the contactual stage in the third trimester, when the folds of the fetus's body against themselves lead to excitation at that level of development, causing the later adult to seek pleasure through sensual contact. This stage is augmented by a construdo or self-connecting construdo stage of combining in utero reinforcing this as its aim to combine with things.

In summary, there are two drives of the psyche created by the forces of physics, construdo and destrudo, which operate according to principles of physics, and a third (libido) that operates according to quite different biological principles. Libido, in the unconscious, does not conflict with either construdo or destrudo, although it often joins with construdo, which puts it into conflict with destrudo.

Forming libido during the first six months of uterine life is augmented by self-combining and self-connecting construdo, and the two are in conflict with break-apart self-destrudo. Such conflicts continue throughout a person's life and are reflected in their behavior in all their activities and relationships. This is the life-death conflict. It invades our human activities and enterprises. We build up things in the space of events (the construdo influence) forward in time. We break down things in the space of events – into pieces, so that they can never again exist; or back to the time before the thing existed, backwards in space-time – the destrudo influence.

Pleasure comes from participating in this conflict, as does the feeling of being involved in real life conflicts, because of the contribution and participation of libido, the biological drive; it is what is most biological within us, but the other two drives link us with the forces of the universe. Libido is what Freud (1905/1953b) found to be most human in people as a drive. But this finding is misleading as to the true nature of humans; that is, humans are not essentially biological creatures. Rather, they are the living expression of forces of physics from the universe. Thus TDT proposed that these forces of physics are the root causes of all human problems. The interaction of the two physics drives (construdo and destrudo) in the human psyche circumscribes the progress or lack of progress in human existence. We either go forward in time (inherent in construdo) or backward in time (inherent in destrudo).

The conflict with time and its direction forward or backward, as well as the thought content (space), derived from a drive's energy or the space it occupies in terms of what we think our lives are about, is a unit of physics, or a new way of looking at human events referred to by physicists as space-time. Conflicts in the unconscious between construdo, libido, and destrudo, and multiple derivatives of the three drives are cast into manifold conflicting approaches in all the enterprises of human beings.

This is seen in physics in the way virtual particles annihilate each other. These actions of the subatomic particles reveal the changing identities of particles through space-time. TDT proposes that's the same way the energies of the drives operate in the human psyche.

Freud could not incorporate these issues of space-time within the drives, because such concepts were not known in his era. Libido, the biological drive, does not embrace the place that is forward in time, or the place that is backwards in time, as an issue in Freud's drive energy considerations. Libido does, however, direct stages of development to a space in life's events and cycles and dormant times in human life cycles.

Freud (1930/1961a) cast libido against destrudo in *Civilization and Its Discontents*. He saw their interaction as the essence of civilized progress, maintenance, and deterioration. He thought destrudo had to be restrained to achieve progress. TDT does not. Rather, destrudo influences should be downplayed or held to a minimum. TDT thinks construdo must be strengthened and its aims focused on to achieve progress – a different view.

Freud saw libido as creating the family, the basic unit of society. He considered people joining each other for a common goal of civilization as reflecting the binding influence of libido, "inhibited aim" libido, combining without a sexual aim. TDT sees the joining of people into larger and larger units – families, communities, nations – as reflecting primarily the construdo influence. Further, the tendency in biology to combine into larger and larger units, which Freud saw as stemming from libido's influences, TDT sees as inherent in the nature of construdo.

Libido recreates human beings and is essential for human existence. However, what humans do after they come into existence is a matter of construdo and libido influences, as well as their interaction with destrudo. Construdo plus libido builds societies and civilizations. Both are necessary for civilization to exist. The conflict in civilizations is between construdo, plus libido, versus destrudo.

Without the construdo drive in his concepts, Freud (1930/1961a) saw only a pessimistic future for human beings. This means he saw libido as no stronger than destrudo. TDT endorses his view in its assessment of the relative strengths of these two drives. After all, Freud originally conceived them, and his conceptualizations were accurate as far as they took his theory. With the conceptualization of construdo and the changed relationships between the three drives, there is a fundamentally new view of human life and any meaning that construdo might contribute. The drives' effects on human beings, on their couplings, friendships, families, societies, countries, and their civilizations, can be seen as stemming from the influence of the three drives, not the two drives presented by Freud.

Construdo is more powerful than destrudo and libido combined. This assertion, under normal circumstances, introduces a new psycho-analytic drive theory, accounting for the ostensible constructive progress civilizations have made. Freud's concepts lack this increased advantage of understanding base upon TDT's (three drives) augmentation. TDT based on physics seems to be more optimistic than Freud's dual drive theory based solely on biology. Humans have existed thus far; triadic theory predicts we will positively, constructively continue to exist until forces of the universe, the forces of physics, cause our end.

Adding construdo to Freudian theory restores hope for the evolutionary human experiment. Freud conceived of libido as the life drive, including in its aims all that is done in life to continue itself, other than procreation. The latter reflects libido, the former, construdo. This means that construdo is the life drive and libido generates the aim of life to continue itself.

Yet understanding our universe – by looking at the effects of its forces on us – continues undaunted. Freud never felt such considerations, because in his time, when biological considerations of how human beings were linked to other living organisms were the focus of intellectual considerations related to Darwin's (1859/1972) theory of evolution. In the mean time, with respect to contemporary thinking in this part of the early 21st century, intellectual considerations have shifted to wanting to understand our relationship with the rest of the universe. It is true that human understanding of the relationship with other living creatures on earth began when Darwin introduced his theory of evolution. The next question that began to cross our minds was how we were related to the entire universe. The present work and the drives of the psyche presented in it may begin to generate an answer to such a question. This question and answer begins with Freud's linking of a drive of the psyche with living creatures on our planet, namely libido. He thought the destructive drive must be similarly linked. TDT's theory of destrudo refutes that concept and continues with the answer relating to a Triadic Drive Theory (TDT) based in physics and not in biology.

THE DRIVES AND PHYSICS

To summarize, the influences of physics that created destrudo and construdo have existed in the universe since its beginning – 13.8 billion years ago. The forces of physics that created these drives do not require life in any particular organisms. Therefore, to extrapolate, it may now become more comprehensible to see that forces of nature can possibly influence the developing brain and body of the embryo and the fetus even before birth and that this influence can continue as long as the human being is alive. In addition, after death, the influences, as pointed out by TDT continue in the dead body.

Freud (1915/1957b) observed that the drives often tend to reverse themselves. So did Anna Freud (1946). They did not explain this phenomenon, but they reported it. Nevertheless, the shift to physics permits us to look for a different explanation of the observable reversal in the direction of drive energies. Earlier, it was indicated that in the forces of subatomic physics, we can see the transmutation of neutrons into protons by the weak nuclear force (Davies, 1982) – representing changes in particle identity (Chester, 1978). Subatomic particles also contain particles, quarks and antiquarks, neutrinos and antineutrinos.

These virtual particles annihilate each other when combined, creating new particles. These changes in energy from subatomic physics underlie and can be a model (a direct relation) in the reversals in the unconscious from destrudo to construdo. Figure 1, below, attempts to show these energy interactions.

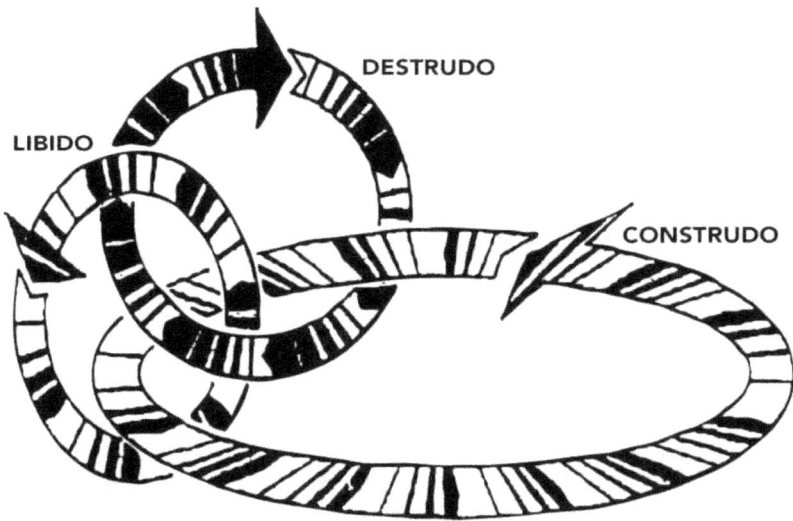

Figure 1.Triadic drive energy interactions.

CHAPTER 9

THE UNCONSCIOUS AND THE SUBATOMIC BRAIN FIELD: FREUD'S DREAM DISCOVERIES OF THE UNCONSCIOUS

CREATION OF THE UNCONSCIOUS BY THE SUBATOMIC BRAIN FIELD AND THE INFLUENCE OF CONSTRUDO IN DREAM FORMATION

Dreams, Freud found, were created in the unconscious psyche. He presented his concepts in *The Interpretation of Dreams* (1900/1953a). As stated earlier and as a brief review, Triadic Drive Theory (TDT) proposes that operations of subatomic particles in the atoms of the fetal brain after six months in utero and following birth up to six months of age produce a field of interactive dynamic forces that influence the brain. Thus, in TDT it is assumed that forces create the operations of the unconscious psyche. The forces of physics that produce this influence from within the atoms of the brain are the electromagnetic force, the weak and strong nuclear forces, plus gravity influences from outside the brain.

Based on Darwin's (1859/1972) work Freud assumed that humans were totally biological. He assumed the psychological drives must therefore also be biological. Consequently, Freud encountered some difficulty, because the biological basis for the aggressive drive, which he postulated in his concept of the death instinct, was never corroborated by biology. However, as mentioned earlier, both the weak and strong nuclear forces in the atom were only discovered in physics after Freud's discoveries of dream operations.

In TDT it is proposed that the most significant influence on human behavior is created by forces of physics, and that biology's only influence is in recreating humans. Biology plays no role in the creation of the universe. Rather, the forces of physics have created a universe of over a hundred billion stars in the Earth's galaxy alone. The Sun is only a single star of moderate size in our unimaginably huge galaxy. And our galaxy is only one of a hundred billion stars (Stott & Twist, 1995). Can anyone doubt such a universe in all its vastness is incomprehensible to humans? This vastness in dimensions of the universe was unknown in Freud's time, as were the principles of atomic and subatomic particle operations.

As humans are creatures of the universe, it is likely that human psychic drives are also created by these same forces. We have described how these forces created destrudo and construdo. TDT considers them the

111

fundamental psychological drives behind almost all human behavior. As noted earlier the libido drive is only the human mechanism for perpetuating our species in the universe (a biological phenomenon) while affording us great sensual bodily pleasure, attraction, and release of libidinous energy.

The human unconscious psyche is initiated from one generation to another by the impinging influence of these universal forces of physics on the fetal brain to form the unconscious psyche. It is an influence that begins anew with each new generation, because the influences of physics are omnipresent and eternal, influencing generation after generation in utero. TDT considers the unconscious psyche as an entity in and of itself, determined by the specific forces that created its specific operations. In TDT this distinction is made on the basis of evaluating Piaget's (1954) cognitive psychological concepts of how a child's mind constructs reality from a mind that is in a state of "undifferentiated chaos." How true, yet how false! Piaget saw nothing logical, reasonable, or sensible in the child's mind before it begins to construct reality. While this so-called primitive state of mind may differ from conscious thinking, it is not proven without organization or structure. Freud's work on dreams half a century earlier than the work by Piaget showed that the unconscious psyche had structure and operational principles of its own. But cognitive psychologists like Piaget ignored Freud's findings. TDT asserts that our unconscious is governed and directed by principles of operation entirely of their own making that have nothing to do with our conscious thinking.

Freud (1900/1953a) outlined the operational mechanisms of the unconscious psyche, which were unknown before his publication. Building on Freud, we assert why and how the unconscious psyche can come into existence and why it operates as it does. Freud said that the unconscious was completely formed when the infant first begins to interact with reality. We concur with Freud because the forces of physics are constant. Hence, they are present in the atoms of the fetal brain in utero, as well as in the infant following birth. While these influences continue, new influences from external reality after birth stimulate the perceptual senses. These outside influences conflict with the internal influences that earlier dominated the fetal unconscious. This conflict between reality perceptions that permit feelings of stability and absence of anxiety, versus feelings of displeasure, instability, changeability, and anxiety about the inconsistency or perceptions engendered by the unconscious psyche, causes an infant to block out perceptions from its unconscious reality, simultaneously repressing these perceptionsand also its unconscious psyche. Of course a newborn would dread returning to such disconcerting, uncomfortable perceptions. Forever after, the unconscious psyche is subject to repression.

During the first few months after birth, an infant's perceptual apparatus, as well as his ability to perceive reality – develop biologically. The infant begins to perceive discrepancies between its unconscious psyche's perceptions of reality, which shift and change before its own eyes, and its external reality, which begins to maintain a persistent perceptual constancy.

In utero, images that were beyond the fetus's control appeared to and disappeared from the visual brain, caused, it is seen by TDT, by subatomic field activities and influence on the unconscious psyche. This condition of images appearing and disappearing is accomplished by the interchange of particles yet to be discovered by biophysics. These maladaptive perceptions present themselves again and again to the infant. Gradually, the newborn will register this conclusion and begin to seek an impression of reality for a perceptual apparatus uninfluenced by the perceptions of its unconscious psyche.

Out of these conditions of maladaptive perceptions being replaced by unconscious-free perceptions, the infant's ego development begins, while seeking to adapt and bind to its new world's conditions of such unconscious-free perceptions. As this development proceeds, the infant's unconscious psyche is repressed further and further, because its perceptions are so unstable, giving the infant neither pleasure nor peace. Everyone in later life resists returning to this shaky world of unconscious perceptions. These are psychotic perceptions from a psychotic world of unconscious perceptions. In its innate wisdom, the infant seeks to find stability in its beginning ego perceptions in order to avoid such a psychotic world. This chaotic developmental stage is described by Piaget (1954) in his *The Construction of Reality in the Child.*

Freud (1923/1961c) declared that the ego develops out of the unconscious, which is greater and more extensive than the ego. He called early repression "primary repression," because the repressed material was never conscious. "Secondary repression" is material once conscious, then subsequently repressed into the unconscious. Thus the ego develops out of and away from the maladaptive perceptions of the unconscious psyche in the infant's new world, also a coincidental difference of unconscious and conscious construdo.

The strong and weak nuclear forces and the electromagnetic force within the atoms (those that create the field of forces in the unconscious psyche)do not require a high level of psychic organization.They can easily influence the fetal or newborn's brain. The brain is formed in the fetus by the sixth month. Its cells are composed of molecules made up of atoms. Within these atoms, particles interact in defined ways. These interactions occur within the atoms of the millions of cells in the fetal brain. During this period in utero these are the primary mental energy interactions in the fetal brain. What the patterns of dynamic subatomic-particle interaction will be can only be stated as probabilities.

Accordingly, in terms of quantum theory, these particle interact-tions set up field forces. These primary interchanges of subatomic particles within the brain cells give rise to mental energies. It's what occurs in the unconscious. Psychology can never predict all the unconsciously motivated outcomes in behavior, any more than physicists can predict the position and velocity of a subatomic particle at any given time. That's the uncertainty principle of physics.

The ego develops from the unchecked impulses of the unconscious psyche. It disconnects itself from these unchecked unconscious impulses of construdo and destrudo and brings them into unconsciousness. TDT conceives conscious construdo much as traditional psychoanalysis conceives the ego.

Consider for a moment the panic or acute anxiety states that some people experience when their primary repressions threaten to break through their primary repressive barrier. The person fears becoming psychotic, because returning to the perceptions of the unconscious psyche would mean going back to an unconscious psychotic world.

Furthermore, the imagery around which some dreams are formed suggests a return to psychotic world perceptions. Dreams that occur when one has a high fever demonstrate the point. In these dreams, one steps down from a high plane to a lower plane, but the new lower plane alters when one steps on it. It grows wider and longer. All reality is constantly changing. Such dream images relate to psychotic world perceptions stored in the unconscious psyche. A patient described this type of dream as being in a maze he was trying to escape, but the walls of the maze kept changing and shifting.

The unconscious psyche is the source of destrudo, construdo, and libido, as well as the source of human creativity. It is also the source of sleep dreams and daydreams and extrasensory perceptions. The uncon-scious can solve problems when egos are blocked. Freud (1900/1953a) noted the potential of the unconscious's creative abilities to solve problems, as when one goes to sleep thinking about an unresolved problem and awakens with the solution.

Permitting the unconscious to solve reality problems is usually resisted by the ego because it has already completed its repression of the unconscious. To the ego, unconscious thinking is too destabilizing, too changeable, too disconcerting and unpleasurable because it is so anxiety provoking.

The ego's conflict with the unconscious psyche, which led to the latter's repression, need not be considered proof that the unconscious psyche has no value to the ego. After all, the unconscious is capable of creative solutions – solutions born in the unconscious. Thus the unconscious remains a source of untapped resources and solutions to the problems of conscious living, because some problems are solved in the unconscious, when in fact it was unconscious obstacles that caused them in the first place.

Just as a dream is formed from unconscious drive, one's thoughts present themselves in the manifest dream content as mysteries for conscious understanding; so too unconscious emotional motives are disguised in the manifest dream content. Moreover, solutions that the creative unconscious might devise to conscious problems will often be consciously resisted, because their origin in the mind cannot be understood.

Humans are afraid of returning to the perceptions of the unconscious psyche because they have been found by the ego to be radically out of touch with the environment and specifically out of touch with reality. But this does not mean that everything in the unconscious psyche is unrealistic or cannot solve reality problems. Creative artists, poets, composers, sculptors, and writers attest to that assertion. Their work or creations come from the unconscious. The unconscious mental energy forces of physics create destrudo and construdo, which connect humans with reality. Hence, two thirds of the human drive energy is contributed by forces of physics, not biology. The unconscious stores the deepest human knowledge of earth and the universe, because according to Triadic Drive Theory (TDT) the unconscious psyche was created by the same forces that created the atom and the universe. Additionally, it is considered by TDT that the force of gravity holds the universe together and contributes to the unconscious's creation of construdo.

It is interesting to see that the essential, structural characteristics of dreams as discovered by Freud compares closely to the physics of the subatomic particles operations in the atoms of our brains in the sense of corresponding with Freud's description of the nature of unconscious mental operations.

FREUD'S TIMELESS AND SPACELESS DYNAMIC
UNCONSCIOUS AND THE INTERIOR OF THE ATOM

A primary difference Freud found between the nature of the dream world and the world of reality was that dreams did not adhere to time as we know it consciously. Representation of time is often reversed in dreams. The beginning of manifest dream content may represent consequences of a later dream thought. A dear, loved person who has died in reality can appear alive in a dream. Feelings from the distant past can appear vividly in a dream with an intensity that makes the dreamer feel they had occurred only yesterday. Freud (1900/1953a) saw the unconscious as timeless (e.g., not having the time reference we hold in our everyday reality). Subatomic particles can change identity as they move through time. Clearly, within the atom, time does not exist as we know it. Similarly, some people regress to earlier emotional states of their existence where the infantile needs of their past can determine their current adult behavior.

Timelessness of the Dream

Freud's fundamental finding of the transformation of dream thoughts as they appear in manifest dream content (what we remember and report as the dream we had last night) has, theoretically, a counterpart in the operations of the subatomic particles within the brain's atoms. Physicists call this *transmutation* of particles. Furthermore, the electron and positron are antiparticles of each other. If they collide within the atom, they annihilate each other, and this produces two or three new particles (Davies, 1978; 1982). Freud's transformation of dream thoughts into manifest dream images being distorted from true latent dream thoughts is analogous to such operations of physics.

Displacements in the Dream Work

Another comparison between dream work and subatomic particle operations is that the position and velocity of a particle within the atom can never be known exactly at the same time. The more precisely one factor is fixed, the less precisely the other can be determined. Moreover, quantum mechanics, which was combined with special relativity theory to explain operations of subatomic particles, indicates the probability of a particle's positron can only be predicted when its speed is measured exactly. However, this cannot be done, because such measurement distorts the measurement of the position or the place of the particle. We are left with the uncertainty principle.

Similarly, in the operations of the unconscious in dream work, one can never be certain where a dream thought will emerge in the remembered dream. What Freud called "dream work," TDT sees as the field of forces created by the operations of particles within the atom and their exchanges of energies within the brain. In fact, dream thoughts, Freud found, move from one likely place to another less likely place and do not maintain the spatial location that conscious logic and reason would dictate. Freud called these distortions "displacements." Ideas, affects, impressions, or images of the dream thoughts could move in the dream work to the least likely place in the manifest content. The former impressions or images are thus disguised. A dream image at one place in the manifest dream can represent an idea, affect, impression, or image derived from or distorted from a very different place in the dream thoughts.

For example, dreaming of an action, feeling, or idea related to the mouth or eyes could represent actions, feelings, or ideas related to the genitals. That is a displacement upward from the genitals to the head. Central dream thoughts of high intensity value are often not apparent in the manifest dream. Reality, Freud said, is disguised by the appearance of a seemingly unimportant dream part. He said this was because of a "displacement" of central dream thought to some other, seemingly minor part of manifest dream content (Freud, 1900/1953a).

Spacelessness of the Dream

In dream work positions of reality, relationships, Freud said, do not exist as they do in reality. Chester (1978) pointed out that subatomic particles behave without reference to time or space. Importantly, with respect to Triadic Drive Theory (TDT), Chester (1978) noted that only in dreams is there a parallel type of thought as there is in the operations of the subatomic particles within the atom.

Condensation in the Dream Work

Freud found another essential characteristic of the dream work. It was what he termed "condensation." Dream thoughts are regularly and characteristically compressed into smaller and smaller dream images in the manifest dream content. This process is parallel to the processes in subatomic particles operations called "confinement" in physics.

The nucleus of the atom is composed of protons and neutrons, the result of combinations of quarks. The smaller these subatomic particles become, the greater the energy required to compress them or hold them together (Hawking, 1988). This subatomic property of confinement in the atom binds particles together into righter and tighter combinations (Hawking, 1988). This "confinement" phenomenon is the result of the strong nuclear force (Hawking, 1988). When compressed, combining triplet quark particles produces a proton or a neutron. The implications of this principle from physics will be discussed later in terms of the principle of construdo. Subatomic particle investigations have found smaller and smaller subatomic particles compressed into each other as are those composing the atomic nucleus. Current string theory is looking into what is the smallest unit, possibly a string that is uniting the forces of physics.

This transformational process could easily be seen as an influence of the construdo drive and the confinement influence in the dream formation causing transformation of the latent dream content into the manifest dream content. Thus dream thoughts are combined and compressed into smaller dream images. These dream images go forward in time under the construdo influence and its direction in combining images of the dream. Still, the dream process is also influenced by destrudo and its backward direction toward earlier construdo conflicts as well as libido conflicts with destrudo. All these drive influences and conflicts produce dream image transformations. Destrudo takes the latent dream content backward to earlier times of conflicts. Often, they are earlier libido conflicts.

Freud (1900/1953a) indicated that the basic dream transformation processes of displacement and condensation can combine, producing even more disguised, more succinct manifest dream content. In total, TDT posits that the operation of subatomic particles and their property of confinement is the underlying basis from the forces of physics within the atom that causes dream thoughts to pull together in complete construdo fusions.

Moreover, Freud indicates that the resistance of conscious forces later in life can combine with unconscious drive impulses seeking expression through entrance into consciousness, resulting in compromise between construdo and destrudo. Construdo also combines these two (the resistance of conscious forces and unconscious drive impulses) into a compromise formation, which results in the repressing superego accepting a disguised version of these drive impulses.

Freud further indicated that overcoming the superego censorship, or resistant conscious forces, is often accomplished by distortions created by the dream work. He noted that when a dreamer thinks, in the middle of a distressing dream, "[t]his is only a dream." This indicates that the dream has developed beyond the censorship. The "censorship has not been prepared to censor what is occurring in the dream." The words "this is only a dream" dispense with the anxiety and distressing feelings aroused by the dream. This is part of the secondary revision influence from conscious thinking (Freud, 1900/1953a).

My mother dreamed about her own impending death; and she said to herself in the dream, "To hell with this, I'm waking up." She did. That is often the result of such distressing dream thoughts.

DISTORTIONS OF THE DREAM WORK AND THE PHYSICS OF THE DYNAMIC OPERATIONS OF SUBATOMIC PARTICLES WITHIN THE BRAIN'S ATOMS

Freud believed that radical changes from the dream thoughts to manifest dream image's demonstrated "distortion" of the dream thoughts. When unconscious dream thoughts form a dream, none of these thoughts retain their original value. They are moved, shifted, or displaced to images in the manifest dream content that may even, or especially, appear trivial. When a dream appears to be confused and obscure, Freud found that such a dream was constructed with marked displacements from the dream thoughts. Regression of the dream thoughts into deeper levels of the unconscious psyche leads to even more disguised or distorted perceptions of the dream's true meaning (Freud, 1900/1953a).

In TDT thinking these distortions of the dream thoughts into manifest dream contents are the result of the multiple effects subatomic particles have on each other. We know from physics that to achieve such interchanges of energy, there has to be an interchange of subatomic particles or sub-subatomic particles between two sources of energy. In this instance the exact particle has yet to be discovered in biophysics of the brain. Until TDT, it appears that there has been no reason to search for such an exact particle. These interactions are responsible for the way the derived unconscious psyche operates, that is derived from these sub-subatomic particles' interactions.

Subatomic, sub-subatomic particles, and the unconscious psyche's creation of dreams have a logic and operation principles of their own; very different from our conscious logic and from any conscious understanding of our reality.

Significantly, the logic derived from the mechanistic physics of Isaac Newton (1687/1962), the physics of conscious, scientific cause-and-effect thinking (and not the physics of systems of dynamic interactions), is more in line with conscious reality thought than unconscious dynamic reality, where a change in one part of the system results in subsequent changes in all other parts of the system. Newton's (as well as Freud's) is the physics and psychology of determinism, the science of the 19th century.

Freud's findings and assertions about the unconscious are more similar to the findings of modem subatomic physics regarding the workings within the atom. As indicated, Freud, however, imposed the doctrine of determinism on psychoanalysis, making it a part of the deterministic science of his time. He introduced his concept of a dynamic unconscious in 1900. This concept of a dynamic unconscious actually corresponded to the physics of subatomic particles that would not be discovered in physics until a quarter of a century later. .

In present day physics, understanding of operations of subatomic particles within the atom stems from quantum mechanics and from Einstein's special theory of relativity. These factors in TDT establish physics based underpinning in understanding the underlying, strange, illogical operations of Freud's unconscious psyche's operations.

The transmutation of particle physics(when a subatomic particle passes through time, forward in time, then backward in time), is difficult for conscious minds and logic to understand. The concept defies conscious reality perception. Ordinarily, everyone takes for granted that time can only move forward. However, this conscious belief about time has been questioned in modem physics (Chester, 1978). Freud not only saw in conscious mental processes regressions to earlier emotional feelings and patterns of reaction but also observed regressions or a movement backward in time in the dream work of the unconscious psyche.

The backward-in-time energy is supplied in a dream by destrudo. The dream image can regress to a space-time where libido was tied up or encountered difficulty to enhance forward development, or to a point where destrudo encountered difficulty to support further forward development. The dream's forward direction in space-time energy indicates an earlier wish to overcome these regression difficulties. Such a wish represents an attempt to move forward again if these libido and destrudo obstacles could be removed. If so, the individual's construdo can get the person moving forward again. Thus, and perhaps not so surprisingly, because of Triadic Drive Theory, dreams can tell psychoanalysts why individuals cannot move forward in their lives.

A dream reveals a past wish of construdo that was blocked or frustrated. Freud's (1900/1953a) dictum that a dream represents a wish did not conceive of construdo (nor that the wish might refer to past wishes of construdo), a drive whose aims are future wishes. Hence, the wish component might refer to past frustrated wishes that continued into present times. The destrudo component of the dream took the dream thoughts backward in time to where it was aroused, inasmuch as these construdo wishes, anticipations, or expectations were blocked, inhibited, or frustrated. These blockages might originate in the libido, destrudo, or construdo. The dream wish that Freud reports to solve a present drive conflict refers to earlier times of the conflict, and the blockage – to solutions, then and now.

The concept of space-time from physics (Hawking, 1988) allows us to view what an idea in conflict is (space) and when it becomes a problem (time). This is a concept for psychoanalytic diagnosis, for formulating the problem and its treatment. Space-time a concept from modem physics that well fits the concepts from which psychoanalysis stems. That is, past time frustrations cause new problems in individual's present lives. Indeed, the distortions of a dream-thought find correspondence in the dreamer's past, causing creation of the dream.

Freud asserted that the unconscious was in no way governed by conscious concepts of time. Rather, the unconscious produces gross time distortions in the manifest dream content. The regressive influence in the unconscious parallels destrudo's influence to break down or apart what has existed but is not wanted. According to TDT this regression is derived from the subatomic particle movements backward in time. This regression of dream thoughts into the unconscious's earliest thinking or organizations will lead to a more disguised or distorted production of the dream (Freud, 1900/1953a).

The unconscious is characteristic of the organizational mentality of the fetus after six months in utero and of the newborn's first few months following birth. Regression to an early unconscious organization leads back to these times. Nonetheless, as earlier stated, the unconscious perceptions of reality in the infant's new world prove to be so changeable, so inconsistent, and so unpredictable, that these unconscious perceptions of reality provoke anxiety about what will happen next. The resultant feelings are psychic and emotional experiences of great unpleasantness. Individuals strive to avoid this unpleasantness. Freud asserted this concept a year after his *The Interpretation of Dreams*, in *The Psychopathology of Everyday Life* (1901/1960).

The ego strives to keep the unconscious unconscious. TDT posits that the reason for this is that following birth, the infant is constantly challenged by the new world and finds that its unconscious psyche's perceptions of that world neither lead to nor lend themselves to any consistent, stable perceptions of reality. The ego seeks a way of seeing the world that will prove constant, unchangeable, and predictable.

This unconscious perceptual world is described in Piaget's (1954) impressions of the world in which the infant moves (in his *Construction of Reality in the Child*). He describes it as a world of perceptual "undifferentiated chaos." From Freud's (1900/1953a) assertion that the ego develops out of the unconscious, it follows that the ego's perceptions develop out of the id's unconscious perceptions. Such perceptions seek constancy and predictability from construdo's overriding aim to connect the individual with reality, once in it. It's the tendency Freud (1911/1958b) observed in what he called the "reality principle."

Freud (1900/1953a) indicated that dreams perpetuate the conflicts in a person's waking life as stimulated by the previous day's events. What may consciously be dismissed as trivial in a dream may reflect more of these conflicts than first realized. The dream may relate to many significant experiences in the dreamer's past life. The frustration of one of the drive's intents may cause a persistent drive wish that will not go away.

Distortions of these manifest dream contents are stimulated according to TDT by the multiple effects subatomic particles have on each other in the atoms of the brain, and in turn in the brain's field of forces. The fundamental principles of the unconscious psyche and its operations rest on subatomic particles' operations within the atoms of the brain. This occurs before any biological principle and operations come into existence in the unconscious.

The unifying universal force postulated and sought after in string theory existed before the four differentiated forces came into existence in our universe. It is prior to life, and constitutes the basic psychic drives in all living things and of unconscious energies. Thus, such a universal force is a force not pertinent as an influence on human beings or any other living things on our planet.

Freud's Concepts of Dream Formation and Construdo

In Freud's concepts of dream formation by the dream work, there are four stages comparable to the four stages of construdo.

First, ideas, affects, impressions and imagery of the dream thoughts are assembled. Freud says they are combined around a "nucleus" of the dream, attracting the material of the dream thoughts around itself. The condensation processes combine more and more of the dream thoughts into smaller and tighter units.

The logical relationships, on which the dream thoughts are combined, according to Freud, are similarity, consonance or harmony of sounds, as well as common attributes. He said that one and only one factor will be responsible in a given dream for combining the dream thoughts. Remember that combining with what "surrounds" is the first stage aim of self-construdo in utero and of object construdo after birth.

Second, dream thoughts are connected with the previous days' residues, thoughts, affects, impressions and imagery; that is, they are connected to any conflict the individual is experiencing in the present but has not solved or resolved when these drive conflicts first occurred. A solution to this conflict is contained in the fulfilled wish of the forming dream, which solves the earlier drive conflict problem. The connecting dream thoughts are transformed and disguised in the manifest dream content or in the intermediate associational linkages in dream work.

Freud's associational method, of underlying dream interpretation, permitted latent dream thoughts – the real, underlying thoughts causing the dream – to be uncovered. This connecting aim is that of the second stage of both self- and object-connecting construdo. Thus, the formation phases of dreams are processed as construdo imagery progresses in utero and following birth, during the first few months. The dream formation processes are primitive because they begin in utero. They pull together, or combine by similarity of sound, common attributes, and we add proximity as a basis of unconscious dream formation. Dream formation continues until the symbolization processes begin after birth, when the infant connects the ideas that verbal symbols stand for things, at about one year of age.

Third, Freud said that the unconscious psyche in the dream work constructs a dream out of the created intermediate dream images. A plot for these combined and connected dream thoughts in the intermediate asso-ciational images is constructed by the unconscious psyche. The progression of the dream by association is the result of the mental characteristics of the unconscious psyche's perceptions operating on reality and being affected by that reality. Yet we can sense how different a consistent reality is from that of the constantly changeable unconscious. Rather than being deter-mined solely by the operations of subatomic particles within the brain's atoms, after birth the unconscious combines with reality and seeks to find itself in conscious perceptions of the ego. The constructing, forward-moving-in-time stage aim of object construdo is responsible for this third stage of dream formation.

Fourth, creative dramatization of intermediate dream thought material embellishes its composition, possibly to entertain the dreamer. This is suggested by "ingenious," "symbolized," "composite" figures, "displaced" to other figures creating the manifest dream content (Freud, 1900/1953a). The creative embellishment of dream symbolization, the final phase of dream formation, is tantamount to the fourth object – construdo stage, the creative stage. This is what Freud defined as "secondary elabo-ration."

TDT therefore has reached the conclusion that dreams are the product of construdo's workings. The construdo drive combines and connects the stuff dreams are made of, and then constructs and creates a dream.

Representation in Dreams

All dream transformations must be represented in pictorial form and sequence in visual language. Freud noted that one of the favored means of pictorial transformation in dreams is by turning an idea, affect, or image into its opposite or in TDT physics terms, an antiparticle. This operation underlies the reversing dream work operation, specifically the tendency of the unconscious psyche to reverse itself.

Contradiction of ideas or affects in dreams is the dream's way of saying "no" pictorially, Freud asserted. Opposite thoughts may be represented pictorially in a dream through displacement, when some part of the manifest dream content is turned into its opposite, appearing as an idle thought. Inhibition of the movement of a dream character can represent a conflict between the id or unconscious impulses of the dreamer and the conscious ego's desires. An example: one dream character wants to flee an approaching threatening character, but is unable to move. Absurdity or apparent nonsense in the manifest dream content can also indicate contradiction of an idea.

These twists in dream formation are consequences of time having no meaning when negation of an idea underlies the unconscious aim forming a dream. Secondary revision in the manifest content to some degree tries to take time into account. Another example of the time difference between unconscious dream formation and secondary revision in the manifest content is seen when Freud indicated that alternative views in dream thoughts cannot be represented in the dream itself. When the dreamer reports, "This was this; no, it was this," Freud (1900/1953a) said the interpretation should proceed as if the dreamer had said, "it was this, and it was this."

The problem is that conscious time logic tells us two different things cannot be true at the same time. This contradiction does not apply in the unconscious, where contradictions can be valid; Thus, conflicting alternatives might both be valid, apart from the conditions of time. The unconscious is not affected by contradictions and alternatives because conscious reality time is not present in the unconscious.

TIME AND PHYSICS

In line with this distinction, physicist Stephen Hawking (1988), in *A Brief History of Time*, asked a question indicating that time was not an absolute under all conditions: "Why do we remember the past but not the future?" (p. 144). Hawking believes the answer is that humans exist within an expanding universe. If the universe were contracting or collapsing into itself, then people would remember the future, not the past.

Hawking (1988) explained that time has a direction, as indicated in the second law of thermodynamics. It states that within a closed system things move from order to disorder with the passage of time. This is the principle of entropy. It increases with time. This principle confirms human conscious observation and knowledge of the passage of time. We know as time passes, things get more and more disordered. Psychologically, Hawking said, people understand this principle as their sensation of "time passing." Time does so in the same direction in which the universe is expanding and continues to move in the direction of an ordered universe that remains governed by the laws of physics. With this universe, however, things can get disordered in time. Consider what would happen if you were away from your office for a year. Would things not be more disordered when you returned?

Hawking (1988) asserted that ultimately, in billions and billions of years, the universe will begin to contract and eventually collapse into itself. The black holes in the universe are stars that have already collapsed into themselves. Their gravitational force is so intense that their gravity has pulled the star back into itself. TDT predicts that collapse of the universe will follow a similar pattern. Things will then move from a state of disorder more and more toward a state of order. Then people will experience time backward, "remembering" the future not the past (Hawking, 1988).

REPRESENTATION THROUGH SYMBOLISM

Among the indirect methods of representation, Freud (1900/1953a) listed the existence of symbols in dreams:

> In the case of a symbolic dream – interpretation of the key to the symbolization is arbitrarily chosen by the interpreter: whereas in our cases of verbal disguise, the keys are generally known and laid down by firmly established linguistic usage. (pp. 341–342)
> [Further], there is no necessity to assume that any peculiar symbolizing activity of the mind is operating in the dream-work, but that dreams make use of any symbolization which are already present in the unconscious thinking, because they fit in better with the requirements of dream "construction" on account of their representability and also because as a rule they escape censorship. (p. 349)

Once a dream interpreter has the right idea, these symbols can be determined without the dreamer's contributing further information. Dream symbols are among the indirect means of representation. Examples of symbolism are also provided by Freud (1900/1953a): "The Emperor and Empress (or the King and Queen) as a rule really represent the dreamer's parents: and a Prince or Princess represents the dreamer himself or herself" (p. 353).

125

In some cases, there is obviously a common element between the symbol and what it represents; while in others the relationship is more puzzling (Freud, 1900/1953a, p. 352). The latter (what the symbol represents) reveals a genetic characteristic. Freud had in mind an erotic reference lost historically in the sexual meaning of certain primal words. These primal words came to be applied to so many other activities and things that the sexual connection was lost. The presence of symbols in dream facilitates their interpretation, while at the same time making interpretations more difficult. The free association method of interpretation leaves the interpreter at sea where symbols are concerned. Freud cautioned against wild or intuited interpretation of symbols; he felt that such a method led to criticism that such wild or intuited interpretations were unscientific. Despite these caveats, Freud (1900/1953a) asserted that:

> ...all elongated objects, such as sticks, tree trunks, and umbrellas (the opening of these being comparable to an erection), may stand for the male organ [added 1909]—as well as all long, sharp weapons, such as knives, daggers and spikes [added 1911]. (p. 354)
> Boxes, cases, chests, cupboards and ovens represent the uterus [added in 1909], and also hollow objects, ships, and vessels of all kinds [added in 1919]. Rooms in dreams are usually women. . . . Stepladders or staircases, or as the case may be, walking up and down them, are representations of the sexual act. (p. 355)
> To represent castration symbolically, the dream work makes use of baldness, hair cutting, falling out of teeth and decapitation. (p. 357)

Contrary to these castration symbols, animals whose pulled-off tails grow back equal to intent of warding off castration, its feelings, and its fears.

Symbolic representations indicate an advance to a higher level in the unconscious organization over that seen in primary organization of the unconscious in utero and during the first few months following birth. These symbolic representations are the product of primary unconscious construdo combined with the infant's unconscious and conscious perceptions of the outside world. After six months of age and beyond, the child learns that symbolization of objects occurs in words or language representing these objects. Construdo in the unconscious constructs symbols for ideas and objects in dreams, then dream language progresses to symbolization. Construdo and destrudo are symbolized in dreams. Symbols of knives, daggers, or sticks, referred to by Freud as phallic symbols, are also symbols of striker-object destrudo. These phallic symbols are combinations of the two drives. Dreams of buildings being destroyed can be symbols of destrudo overcoming the constructing stage aim of construdo. Pictorial representations of construdo may be symbolized by groups of people because they involve combining and connecting with others. Places where one connects and joins the knowledge of the world, like a library or school, may also symbolize the combining and connecting stage aims of construdo.

TIME, SPACE, AND FREUD'S UNCONSCIOUS

For time to be valid as its passage it sensed, Hawking (1988) said, specific conditions must exist. These phenomena of physics have to demonstrate a definite direction that occurs in a given sequence, but not the reverse. The universe must remain in its expansion mode. It is clear from Hawking that time is not an absolute within the universe.

Within the atom, time does not exist in human terms. Subatomic particles can move backward in time, then forward, changing identity as they shift. This is a far different reality from human conscious reality. This moving backward and forward in time by certain particles within the atom is the same operation principle that Freud found occurs in dreams. "A dream might be described as a substitute for an infantile scene modified by being transferred onto a recent experience" (Freud, 1900/1953a, p. 546).

Physicists speculate constantly on the actions of particles and antiparticles, and that there may also be the anti-reality one, composed of antiparticles. It can be deduced that what is known as "reality" depends on the principles of its operation and the conditions that govern such operations. In this volume, TDT posits a dualreality world: one reality from Newton's physics, the world we consciously know, the other from the reality of subatomic particles' operations within the atoms of our brains, our unconscious reality. These are the same two realities proposed by Freud (1900/1953a) in *The Interpretation of Dreams*, only his were distinctions of the mind, and not of different realities from the world of physics.

Time is one element in human conscious reality, spatial relations the other. Freud found neither existed in the unconscious. Physicists find that spatial relationships are maintained within atoms. Theoretically, a particle *can* be in two different places at the same time. Moreover, physicists find space is curved within the universe (Hawking, 1988). The human unconscious operates with full acceptance of spatial displacements that defy the existence of space as a reality. Displacements, Freud found, could move the point of a dream, its central affect or its central conflict, to another place in the dream's manifest content, then on to a completely different dream conflict in some other part of the dream that represents conflict of affect or ideas. The location of the problem leading to the dream is thus uncertain.

It is TDT's contention that Freud discovered the principles of perceiving, thinking, and feeling about a separate reality where humans also exist sometimes, quite apart from their conscious reality existence in the universe. The other reality of our human existence is found in our subatomic particle operations in the atoms of our brains. In dreams, humans visit that reality. Within our unconscious we operate from that reality. In contrast, in conscious reality, humans operate and think within the mechanistic, cause and effect logic of Newtonian physics.

FREUD UNCONSCIOUSLY INTUITS CONSTRUDO

Freud taught us a method for interpreting the unconscious influence underlying an individual's conscious thoughts. Following his concept, we found the first three sections on the dream work in *The Interpretation of Dreams* replete with words related unconsciously to the construdo drive. We ascribe to construdo the psychic energy that assembles all dreams. From Freud's writings relating to the dream work, TDT deduced an underlying process like construdo supplying the psychic energy that pulls a dream together.

Freud often used terms that relate to binding processes in dream formation. Again and again, he used terms such as "common element," "combinations," "composition," "composite figures," "composite," "components," "comprise," and "compose," "connection," "connections," "construct," "constructing," "creating composites," and "creating." He also referred to anticathexis or disconnecting processes, which we consider a fusion of construdo and destrudo processes. It's an example of destrudo overcoming construdo, or breaking apart what has been put together in the unconscious.

Disconnecting construdo can also be the result of construdo's ceasing to combine and connect with certain ideas, affects, or memory images. All these terms refer to binding or uniting processes or unbinding processes. Both lie behind construdo.

These joining or uniting construdo terms describe the four stages' aims in the development of self construdo in utero and object construdo during the first six months following birth and the six ensuing years – the sequence of combining, connecting, constructing, and creating in human construdo development. It is the same sequence of mental processes that form a dream. That is why the Triadic Drive Theory (TDT) proposes that Freud unconsciously intuited a construdo-like process behind dream formation.

CHAPTER 10

FREUD'S "DREAM WORK" AND UNCONSCIOUS CONSTRUDO

"FALSE CONNECTIONS" OF CONSTRUDO REASONING AND FREUD'S SECONDARY REVISION OF DREAMS

Freud (1900/1953a) described how the language of the unconscious is translated into a language the dreamer can remember consciously. This translation occurs when the latent dream content morphs into the manifest dream content by the dream work. In intermediate stages of the dream work, Freud found secondary revision of such unconscious dream work, where secondary thought processes or conscious thinking are driven or transformed to make the unconscious intelligible, reasonable, and logical. This transformation of unconscious material into what is a manifest descriptive story-line is equivalent to the mind's tendency to make logical connections and coordinate dream thoughts logically and reasonably in dream work. This concept runs counter to the real nature of unconscious thinking. The dream work is derivative in terms of visual pictures that wind up in the images we retain in the manifest dream content. We will see that this dream work (transformation process of unconscious material to descriptive story-line) is quite extensive.

Secondary revision is the work of self-construdo influences on primary processes or unconscious organization, a system, in Triadic Drive Theory (TDT) terms created by influences of the operations of subatomic particles within the brain's atoms. Construdo combines and connects these intermediary images of dream thoughts with logical, rational concepts of their meaning and consequences, while the dream moves in a parallel way to construction and creation of the manifest descriptive dream content. This results in two dreams. One is the actual dream, the latent dream content (the dream from below). The second dream is the dream that becomes the remembered dream – the manifest descriptive dream.

The manifest dream makes "false connections" with the logical, rational world with which it is also connected (owing to its experiences in the rational world following birth). Furthermore, object construdo strives to combine and connect with this new world by asserting a rational, logical organization of the intermediary images of the dream thoughts in the manifest dream contents, differing from the latent dream content. The work of self-construdo through its developmental stages of combining, connecting, constructing, and creating, leads the unconscious processes into

129

the new conscious world. This bridges the gap between the unconscious world of self-construdo and the post-birth world of conscious object-construdo desires and motivations.

The energies of both self- and object-construdo stem from the forces of brain operations, whichthemselves, in TDT terms, derive from the operations of subatomic particles within a brain that then provides the psychic energy. This energy derived in the unconscious psyche causes the striving to fulfill the human need to combine, connect, construct, and create a union with things in the world following birth.

The capacity of self-construdo to combine or connect images and ideations in the unconscious psyche (in utero and during the first six months following birth) is seen as Freud's concept of secondary revision in dreams. This potentiality of connectivity stems from the strength of self-construdo and continues, unconsciously, in the object-construdo that also develops during the first six months of infant life. It is at this time that construdo's secondary revision of conscious or secondary process thinking makes the same false connections of logic in conscious thinking that Freud found were referenced by unconscious thought in dreams.

While Freud (1900/1953a) never articulated a construdo-like process, he did note that "secondary revision is the one factor in the dream work which has been observed by the majority of writers on the subject and of which the significance has been appreciated" (p. 501). Rather than leaving the impression that he had completely penetrated the matter, Freud preferred to quote the conceptualizations of others on this problem, because they stated the problem effectively without giving an explanation for it. For example, he quoted Havelock Ellis's *The World of Dreams*:

> We may even imagine the sleeping consciousness as saying to itself in effect: Here comes our master, Waking Consciousness, who attaches such mighty importance to reasons and logic and so forth. Quick! Gather things up, put them in order – any order will do – before he enters to take possession. (Freud, 1900/1953a, p. 501)

In a footnote on the same page, translating the French in his text, Freud (1900/1953a) quoted Delacroix: "This interpretative function is not peculiar to dreams. It is the same work of logical co-ordination which we carry out upon our sensations while we are awake" (p. 501).

Precisely. This is the problem Freud left unresolved – the infra-structure of such secondary revised thinking, which in terms of TDT is explained by the introduction of object-construdo to the idea of a secondary revision of our conscious thinking. The false connections concept that

Freud applied to secondary revision of dreams or unconscious thinking, we apply to conscious thinking. Construdo operates inherently both in its unconscious surroundings and unconsciously in its external, conscious reality world.

Just as Freud perceives secondary revision as resulting in misunderstanding of dream thoughts by false connections, so too, human beings in their conscious thinking arrive at misunderstandings, erroneous connections, or false connections because of construdo's inherent need to link up and bind things within its reality. Consequently, construdo can make false combinations of reality stimuli and can make false connections of reality events, objects, and people. It can also produce false creative perceptions from the realities surrounding it. All this can be the result of an erroneous influence of the unconscious object-construdo drive aims or because ofits four differing stage aims. In such cases of erroneous influence based on the construdo drive aspects of reality are really not related or connected but are observed and assembled by the unconscious as if they were related.

This process of unconscious uniting does not exist in an individual's awareness, but its results do. These results of unconscious uniting are experienced as central, and they are deep within the individual, therefore evoking emotions that support the false connection. Emotions are elicited when one or more of the three basic drives (construdo, destrudo, and libido) is/are involved in an interaction with each other or with the environment. Freud (1966a) first used the term of "erroneous connections" in his letters to Fliess (1892-1899); he continued using it in the *Studies on Hysteria* (1893-1895/1955c) (or neuropsychosis), and in the *Project for a Scientific Psychology* (Freud, 1895/1966b). In all, the unconscious motive is falsely connected to a different motive in the individual's conscious behavior, a transformation of unconscious impulse. In his paper "On the Universal Tendency to Debasement in the Sphere of Love," Freud (1912/1957d) refers to this term, "erroneous connections," for the last time.

CHAPTER 11

PSYCHOPATHOLOGY

TRANSFORMED ID DRIVES DISTORT EGO FUNCTION IN SPACE-TIME

Whenever there are differences in the dimension of space-time between the realities of the universe and the realities in consciousness in the individual's ego for one, two, or three of the drives, psychopathology can arise. It occurs when one or more of the drive impulses is distorted or transformed as it enters the ego from the id in space-time.

Remember, neither space nor time, as Freud (1900/1953a) indicated, exists within the unconscious. The drive impulses originating in the unconscious may undergo transformations and distortions to gain access to the conscious world of the ego's reality and to avoid the forces of repression and suppression. The drives want to combine with the surrounding reality of the new environment following birth, impelled by construdo, which prompts combining with whatever mental factors present themselves to the psyche. Following birth, many new factors integrate with and organize the three drives under the impetus of Construdo, a drive that, according to Triadic Drive Theory (TDT), derived from the universal forces of physics.

Destrudo, also derived from the forces of physics has two developmental stages in utero. This is when biological forces are interacting with the two psychic forces from physics (construdo and destrudo) to produce the human organism. Furthermore, while the uterine biological forces are interacting with construdo and destrudo, construdo goes through four developmental stages in utero. Finally, we assert that libido has two developmental stages in utero, which were not considered by Freud (1905/1953b).

The first is called the forming stage, directed by genetic inheritance that results in biological growth and development during the first six months in utero. The second in-utero stage of libido development is called the contactual stage. It occurs between six and nine months in utero. It is derived from the contact of fetal folds against each other, as well as from the fetus's contact with its own uterine walls. This is the precursor of contactual stroking stimulation of the skin. It is continued in the newborn following birth. The later adult often desires to be touched, stroked, or held. At the same time the oral, anal, urethral, and genital stages Freud (1905/

1953b) described are proceeding. The combining here is under the impetus of construdo to join oneself with what is being sensed in one's sensuality.

Following birth, tissues creating the erotogenic zones are even more sensitive than the tissues creating other areas of the body. These body areas are also focused on in the early care of the newborn. They are biologically critical areas for survival, such as mouth, anus, urethra and genitals. The care of these erotogenic zones establishes an interaction or combining between the caring adults and the infant, which lends a special prominence to these zones within the infant's psyche. These sensations stem from construdo and libido. They combine and connect the individual with its world in noteworthy ways through a contactual sensuality of libido as well. The two drives, construdo and destrudo, in tandem impact the psyche. In addition, during the developmental stages in utero, fusions orinterconnections among the drives originate under the impetus of the connecting aim of the construdo drive.

Destrudo has five developmental stages after birth: the thrasher transitional stage from self- to object-destrudo, and thepincer, toppler, striker, and smasher stages. In utero, destrudo has two stages: break-apart self-destrudo, and defective part causing self-destrudo.

Following birth, construdo has four self and object developmental stages: self- and object-combining, connecting, constructing, and creating. Thus the four object-construdo stages duplicate the identical self-construdo stages in utero, and both self and object stages are seen after birth.

Individuals can be fixated or overly connected by the aim of any of these construdo stages from any of the three drives, construdo included. The individual then unconsciously strives for nothing else other than the drive that is focused. There can be some lag or developmental arrest as when the development of the individual in terms of one of these drives (and its stages) fails to progress forward in space-time. Perhaps the drive is in conflict with its development, with the contrary motives or drive aims. That is when destrudo aims conflict with construdo aims, as often occurs. Or there may have been a trauma in the drive's development (e.g., a destrudo trauma, such as witnessing a violent encounter between the parents).

How does difference in space and time factors create psycho-pathology? Let us consider what these differences might be in space-time factors in terms of the drives. Normality implies the fewest transformations or distortions in space-time dimensions when drive impulses enter the conscious ego's reality. Psychosis implies transformed, distorted, and confused space-time factors in all three drives. Between these two extremes many possibilities exist.

TDT conceives of differences in space-time that are based on the earlier position determined by the unconscious, but also more basically by the operations of subatomic particles in the atoms of the fetal brain, creating a field of similar forces in the adult developing brain. Within the brain's atoms, space-time factors do not exist. But space-time factors do exist in the ego's reality. In reality, time does not exist in terms of past, present, and future; that is to say that the ego's reality understands the universe in terms of Newtonian physics. The space-time dimensions of modern physics after Einstein is the reality of contemporary physics. However, such reality of modern physics it is not the reality of our egos. Instead, here we have an understanding of reality beyond human perception, beyond ego perceptions. In the contrast between the physics of Newton and Einstein, there is a definition of the limits of our perceived reality on one hand, and the reality presented by modem physics (as well as the reality of subatomic particle operations in the atoms in our brains that creates our unconscious).

Modem physics asserts a new dimension of space-time by considering the universe and its forces and the forces within the atom. Both influence what becomes the human unconscious!

Physicist Fritjof Capra found in *The Tao of Physics* (1975) that Eastern philosophers were able to describe the nature of our fundamental reality in terms similar to those that modern physicists use in their explanation of the structure of the atom. These eastern philosophers were embracing the unconscious psyche and its essential nature, paralleling the findings of modern nuclear physics concerning the operations of subatomic particles within the atom, and the atoms of our brains. Hence, these philosophers described a second reality of human existence.

Visual artists can touch such unconscious visions and then return to their conscious ego's perceptions. Most people repress these childhood visions, yet we appreciate the artists who depict these infantile and childhood unconscious visions of our world in their own work. If we did not repress and suppress these sides of ourselves, or our unconscious perceptions, we might see what they see. We can call such artistic vision as regression in the service of the ego.

However, in progressing to six months of age the infant tries to get away from these unstable, changeable, indefinite perceptions from the eyes of the unconscious. The infant tries to locate itself in its new surroundings. At the same time construdo strives to combine and connect the infant with this new reality in terms of space-time perceptions.

One patient could not reconcile her place in space as a wife, mother, and successful businesswoman. She lost herself among these conflicting roles which set off her unconscious's changeability and instability. These conditions of being drawn back into the unconscious's "type" perceptions are so intensely unpleasant that the individual will more than likely repress all perceptions of the surrounding reality especially if we're considering the first six months of life.

Piaget (1954) described this escape from perceptual conditions as undifferentiated chaos, a perceptually unsettling time of the infant's senses. This chaotic reality is blocked out or repressed even as the emerging ego or construdo attempts to achieve a more stable, combined, and connected way in the surrounding world.

The developing object construdo drive is striving to combine with objects, events, affects, and images from the same surroundings, which cannot occur unless its perceptions of the surroundings are stable. Consequently, the reality perceived by the unconscious psyche must be repressed by the developing construdo drive or the conscious ego. The conscious and unconscious perceptual worlds will not coexist.

Healthy development favors the construdo ego because the unconscious psyche's perceptions are so biologically maladaptive for the infant's survival. The unconscious psyche's perceptions were not maladaptive in utero, but become so once birth has occurred. These unconscious perceptions are not part of the world into which the infant has been delivered, but according to TDT belong to the world of the subatomic particle operations causing the electrical charged brain field that exists in utero. These unconscious perceptions conflict immediately with the surrounding Newtonian world's perceptual reality at birth and shortly beyond.

A writer related a description of the unconscious psyche's perceptual reality. He described a time in his life when he did not know where he was or what he was doing; a period when he felt regressed in time. "It was like I was in a maze, where the walls kept changing." The changing of those walls, which would have otherwise been stable boundaries, indicated he was referring to a time and place in life where stable reality boundaries did not exist. What he was describing was how his unconscious psyche perceived the maze.

These perceptions can be found in descriptions of childhood dreams, particularly when a child has a high fever. Children may report dreams that involve stepping from one spatial plane to another, but the last plane changes underfoot in width, length, color, or depth, moving away from the foot stepping on it. In such dreams, children may remember a disconcerting feeling about their place in the world. Such dreams are made up from the child's unconscious perception of reality. Adults can sometimes remember having such dreams when they were sick with fever in childhood.

When unconscious and conscious perceptions of reality jointly determine individuals' responses, conflict results between the two types of perceptions, yielding a compromise resolution that will be psycho-pathological to varying degrees of intensity. The conflict is between the two worlds: those of reason and logic (Newton's world) and those of the nonlogical, irrational, unconscious world of modern subatomic physics that influences the unconscious brain and its perceived reality.

Whenever these differing sense perceptions come into conflict or incongruence, when we are adults – it results in psychopathology. Understanding this psychopathology involves isolating its sources, its influence from space-time, and then determining how such influences have forced these individuals out of balance with the actual reality in which they live. Unscrambling these relationships can lead to an understanding of their psychic pathology.

Development favors construdo or what we have thought of as the ego, because the unconscious psyche's perceptions are now biologically maladaptive in terms of survival. The unconscious psyche's perceptions in utero were not maladaptive, but they are after birth. They are not of the world into which the infant has been delivered. They were of the world of the particle operations of the subatomic brain field existing in utero. They are not of the surrounding Newtonian world or conscious world's perceptual reality.

Understanding such psychopathology involves isolating from where or when the influence is originating. Thus, as stated, unscrambling such relationships will lead to an understanding of the psychic pathology.

CHAPTER 12

DRIVE PATHOLOGY AND HEALTH: THE SPACE-TIME DIMENSION

THE DIMENSION OF SPACE-TIME

When energy from any of the three drives in the unconscious directly enters consciousness untransformed, a drive's energy will be most healthy and most effective. On the other hand, when a drive's energy is diverted into resolving past or future conflicts that differ from the present situation arousing the drive, the drive's energy response will be least effective. A drive's energy should never be impeded or blocked from its arousing stimulus. The drive's energies should be focused on the situation at hand and not be mired in the unconscious past or in imagined future unresolved drive conflicts.

Such situations are events in a person's life in temporal and spatial terms that relate to whatever it is that is arousing the drive. The events in temporal and spatial terms are also the mode by which modern physics considers such events, namely through the reality dimension of space-time (Hawking, 1988). The space-time dimension has four coordinates: three space coordinates and one time coordinate. Thus space-time reality can be considered simultaneously in four dimensions. Physics tells us that this four-dimensional reality is the true reality of our planet (Hawking, 1988).

Hawking indicated, however, that in modern physics neither space nor time is the absolute. Space and time are not measured the same in all situations or under all circumstances. Hawking's findings contradict common sense and human notions of reality. His assertions follow Einstein's fundamental postulate of the theory of relativity, which stated "that the laws of science should be the same for all freely moving observers, no matter what their speed" (Hawking, 1988, p. 20). Thus, the speed of light (186,000 mps) would be the same regardless of the observer's position and speed.

One consequence of relativity theory indicates an equivalence of mass and energy in Einstein's formula, $E = mc^2$. As a consequence of Einstein's relativity theory nothing can exceed the speed of light. Hawking (1988) further indicated, "that time is not completely separate from and independent of space, but is combined with it to form an event called space-time" (p. 23). Space relates to events or incidents in one's life experiences and where such events occurred. Time relates to when these events occurred.

Hawking (1988) posits, moreover, that space and time coordinates are interchangeable. This interchangeability makes the space-time dimension a seemingly and suitable method for defining the coordinates of a psychological event. In some instances, the psychoanalytic notion of transference will also determine the coordinates of a psychological event. For example the space in terms of one's position for the people involved with an individual in the past will also define these coordinates. In other situations, transformations and distortions by the unconscious will determine the coordinates (space-time) of such an event.

Hawking (1988) indicates that an event in the space-time dimension has a past that leads to the present event and a future that is the consequence of such past and present events. He demonstrates this in terms of the light cone from any event. This schema from physics fits well into the explanations offered by psychoanalysis as regards mental events. Freud also considered that the past often determined the present as well as the future of mental events.

Locating a current perception of the ongoing psychological event (its coordinates) and the past events that influenced the present event, along with our perception of it, and locating how the event will be perceived in the future – in terms of these two retrospective perceptions – may require two or three space coordinates. Thus, space-time can be seen as four-dimensional, six-dimensional, or even eight-dimensional. Therefore, this concept allows full conceptualization in psychoanalytic thinking for defining what is being isolated as a psychological event, as determined by four or more space-time coordinates. Establishing such coordinates, to deal with the psychological data with which we might be concerned, is demonstrated in the space-time diagram shown in Figure 2 at the end of this chapter (p.151). Freud never went beyond establishing the time differences a person uses to place an event in its unique time and place, except in regard to how the person was involved, with whom, and under what circumstances. Apparently, time differences are important because the lack of correspondence with what the current reality was and these differences determines, according to Triadic Drive Theory (TDT), the pathological nature of a person's perceptions and responses.

WHY SPACE AND TIME ARE NOT ABSOLUTES

Consider someone lying on a beach looking at a cloudless blue sky. Suppose the person sees two objects in the sky that look like dots. Suppose that both dots are helicopters and they both remain in the same place in the sky, virtually motionless. Then, one helicopter starts to fly away from the other. From the viewpoint of the person on the beach, it

140

cannot be determined which of the two is moving away from the other. Yet what the person sees in conscious reality is that the objects are moving away from each other, because the distance between them is increasing. The person assumes that both objects are moving away from each other in space. In reality, only one is moving.

The objects are perceived this way because the person holds the commonsense belief that there is such a phenomenon as absolute space and absolute time. Paradoxically, the witness cannot see that one object remains at rest because of the fixed concept of absolute space. If the distance between two flying objects is seen to increase, the assumption is that as time passes the objects are simultaneously moving away from each.

Similarly, consider a situation in which one of a twin boards a spaceship that subsequently orbits the earth at near-light speed for a long period of time. When the ship lands, the twin who deplanes will be younger and look younger than the twin who stayed on earth. This is the "twin paradox" in physics. It is due to the fact that as we move from space toward the earth, time speeds up.This is paradoxical only because on earth we believe time is absolute. In the theory of relativity there is no unique absolute time; instead, each individual has his own personal measure of time that depends on where he is and how he is moving (Hawking, 1988, p. 33).

Accordingly, after relativity theory was accepted, time no longer was considered absolute, as commonsense perception suggested. Hawking (1988) asked, why do we remember the past and not the future? His surprising answer: because humans occupy an expanding universe, and because we have other indicators of time's passing in a certain direction. For instance, the concept of entropy indicates that an organized or ordered system will evolve into disorder in time, but that evolution requires some expenditure of energy. Thus the amount of orderly energy decreases with time (the second law of thermodynamics). The entropy of an isolated system always increases, but once the universe reaches the outer limits of its expansion (as Hawking predicts it will), and begins to contract, then humans will "remember" the future, not the past.

This final future ability to remember the future and not the past is a chronic stasis of past and present as well as a "potential" influence on human beings. Thus, this time element of "potential" or "future" is registered in the unconscious psyche, although it has yet to be achieved. The potential influences from the universal expanding forces of physics affect the unconscious. Therefore, properties of physics underlie the mental focus we have on potential or future events.

Fantasy anticipations of how an individual's construdo causes such a person to join or connect with others in a psychological future event can be a source of constant anxiety in neurotics, in particular since such fantasies usually anticipate negative happenings. In normal reactions, fantasies of such events can be a source of future constructive planning. In the personality disorders, such negative fantasies cause mental readjustments in order to avoid such negative anticipated future events. In psychotic individuals, projected future fears will usually be a source of dread.

Some individuals believe they have the ability to foretell the future. This future gazing attests to the influence of the time factor on the human mentality. Such pervasive influence exists because human beings experience an influence from their unconscious psyches to grasp, understand, and predict the future. TDT understands that such an idea of knowing the future is a potentiality derived from universal forces caused by the future potential universal gravitation's pull on the universe's bodies contracting, to create therefore, a contracting universe. Further, TDT understands that this potential future universal force influences the unconscious in terms of conceiving future potentialities, just as Hawking (1988) said would be the case in a contracting universe.

It is part of the influence based on the dynamics of physics that dynamic mentalities are created by our existence in such a universe. Thus, as stated throughout this volume, human mentalities have been created by these universal dynamic forces of physics and reflect their influences, which further have been, apparently via TDT understanding, deposited in the person's unconscious.

The universe never remains static in time, instead it moves constantly toward something different in its dynamic state of existence. This is a reality of our existence in the universe. In this sense the identity of these dynamic universal forces is what is reflected in our mentalities. Thus drive conflicts as well as emotional conflicts follow the form of the conflicts of universal forces (originating from the tightly compacted energy particle that began the universe when it exploded into an expanding universe). Thereafter, these energies always appeared in opposition to each other. The second reality experienced in our unconscious stems from our planet's influences. How we experience this reality bears on how many space-time coordinates are needed to determine an event.

The number of space-time coordinates refers to the number of coordinate points necessary to delimit a beginning event in psychological space-time and its subsequent influences on succeeding space-time events leading up to the current and future space-time events. This concept is most suited for conceiving events in psychoanalysis. The coordinates define the area in a space-time diagram. These coordinates also fit in with the analyst's thinking about what leads up to a particular direction of happenings in an individual's past, present and future.

The space-time dimension can accommodate past transferential happenings in psychoanalysis; that is, accommodating transformations and distortions from the unconscious that have a reference to past space-time. Current space-time can accommodate individuals' future space-time anticipations and fantasies.

The space-time dimension handles the past, affecting the present and future happenings in terms of these influential coordinates – or in human dynamic terms, on psychological factors that must be fixed to locate an event in an individual's life. Moreover, in TDT terms the space-time dimension can incorporate space and time influences from the forces from physics on humans.

With respect to assessing reality the perception of reality depends on how accurately the perceptual system of the ego can define reality. Meanwhile, other forces from physics, from other realities in our lives, also influence our perceptual connections, although we are completely unaware of them. By definition, they're unconscious. This means we are consciously, completely unaware of these influences.

To repeat, the space-time dimension is dynamic. Any change in one part of the system results in concomitant changes in all other parts of the system. What happens today must affect what happens tomorrow. To wit: the backward influences from destrudo will affect the forward influences of construdo. The forward influences of construdo at times carry libido influences, which will then affect the backward influences of destrudo.

Genetically, and according to TDT libido will recreate a new generation of human cells, accomplishing the creation by being combined by construdo. Further, TDT sees libido as able to cause newly created cells to adhere to each other or combine with each other. This, new masses of cells of our bodies come into existence, or are constructed.

PSYCHOPATHOLOGY CAUSED BY THE INFLUENCES OF THREE DISCORDANT REALITIES ON CONSTRUDO AND DESTRUDO IN ONE'S LIFE EVENTS IN SPACE-TIME

As determined from physics, we all exist in three realities. Remember, as earlier stated, this is the reality of operations of particles within the atoms of our brains, the reality of dealing with forces on our planet, and, finally, the reality of our being an integral part of the entire universe. When there are unresolvable different influences from these three realities on the human conscious and unconscious psyche, psychopathology will result. We are considering here an individual's behavior, life events and discordant influences, which all manifest themselves in the space-time dimension. Still, there is the interaction of our two must fundamental psychic drives (construdo and destrudo), as well as their combining and connecting with libido.

First, influences from the reality of existence in the universe include the expanding influence on human drives and emotions toward expansive reactions of a positive, pleasurable variety caused by construdo with respect to its surroundings. At the same time, the gravitational pull of all the universal bodies within a galaxy, and the galaxies on each other, set a direction for destrudo pulling humankind backward into themselves, in a contracting universe, as in our unconscious fantasies and imagination. Under these potential contracting influences in space-time, people experience depressive emotions. They actually become depressed, expanding no longer, because they are contracting into themselves and their surroundings.

Furthermore, in TDT, the influence of stars that is being created throughout the galaxies prompts and promotes the energies of construdo within individuals. Concomitantly, the explosion of stars throughout the universe, and their collapse into themselves as black holes, or the regression backward in time, prompts and promotes the energies of destrudo.

Libido may attach itself by construdo either forward in time, as in a sexual drive towards another, or backward in time, as in dreams that rework sexually repressed wishes and find release of the blocked wish. The volcanic explosions throughout the universe in bodies of different star systems further contribute to destrudo's formative, unconscious energies.

Second, one sense of reality is derived from humanity's perceptions of the planet. As people are held to the earth by gravity, they feel grounded in reality when they see things they commonly believe – the agreed-upon realities of our world. That's the second set of influences on the conscious and unconscious psyche and on the functioning of construdo and destrudo. These "agreed upon realities of our world" are demonstrated by the fact that astronauts have to learn how to move in the weightless environment of their spaceships. Normally, humans sense foundations, roots, bases, and a building of things upward. Construdo thinking builds on itself. Logic creates the assumptions on which derived ideas are based. Feeling "pulled down"is a further contribution to depressed feelings by gravity, which is also a basis for destrudo's taking events backward in space-time. In psychoanalysis the unconscious is conceived as being at a level beneath consciousness.

Nature's destructive energies from space-time events – gales, hurricanes, tornadoes, cyclones or from floods, water erosion of beaches, fierce rainstorms, and the destructive forces of forest fires, volcanic eruptions, or earthquakes – are all planetary influences that set or institutionalize destrudo ideas in the human mind. The regrowth of plant life following such events reinforces and perhaps validates the concept of construdo's rebuilding, or putting things back together, following destructive events. Humans emulate nature when they rebuild what was destroyed.

While these two influences of construdo and destrudo are registered in the unconscious, a third reality also influences the creation of the unconscious: namely, the reality of subatomic particle operations within the atoms of the human brain before birth and thereafter. These operations are spaceless and timeless, which means the influence of construdo and destrudo can move from any place or time in a person's life to another place or time, without that person sensing the slightest disturbance in unconscious thinking. The phenomenon of unconscious influences compresses events by the principle of the confinement influence in the atom causing particles to compact into tighter and tighter units. This confinement influence causes the condensation influence in the unconscious that Freud astutely observed in dreams.

That one subatomic pair can be at two different places at the same time (as found in subatomic physics), can be seen in TDT as causing the displacement process of the unconscious, observed by Freud (1900/1953a) in dreams. Furthermore, the drive aims can reverse themselves as they move into conscious thinking – another "displacement." The aims or thoughts reflecting a dream image can be compressed or condensed into a symbolic expression in a dream symbol. Freud found that many dream images and thoughts were condensed into fusions of several thoughts. Symbolization is a special instance of fusion, but dreams show many instances of fused dream thoughts. Thus, TDT (Triadic Drive Theory) is able to identify processes in physics that have direct relevance even to the psychoanalysis of dreams.

This comparison of physics with psychology is also seen in the universal contracting forces of gravity versus the expanding universal forces from the Big Bang outward that can affect the unconscious to expanding its drives outward; the gravity pull can influence such drives it to contract back into the unconscious. Space-time locates where and with what we are dealing when trying to understand the products in consciousness that result from these influential operations of expanding and contracting, from the subatomic particle reality – our third unconscious reality influence.

The influence of subatomic particles smashing into each other, and then creating new particles, underlies in our unconscious's destrudo as a backward-in-time destructive force, and construdo as a constructive forward-in-time force producing new particles. As earlier stated, in TDT, this underlies creativity in humans.

In the third reality of the subatomic particle influence on the unconscious, as stated earlier, it is found that when the velocity or speed at which a subatomic particle is traveling is known or can be determined, we cannot accurately determine its position in space. Conversely, we cannot determine its velocity or speed in space if we know its position. In TDT understanding, this principle of physics becomes part of the third reality influence imposing itself on our unconscious psyche. Thus, 19th century deterministic view and philosophy of science is replaced here by the 20th century uncertainty principle of modern physics. Coupled with quantum mechanics, it leads to an understanding of a conscious psyche whose roots are difficult to determine if not actually uncertain themselves.

When these three realities affecting the unconscious psyche (atoms in our brains, planetary forces, and integral parts of the entire universe) are in dynamic harmonywith reality as we commonly know it, humans experience mental health in terms of their drives. When the three drives' influences are out of balance or harmony with each other (construdo, destrudo, libido) along with their three unconscious realities (cosmic, planetary, and subatomic), then psychopathology results.

Construdo and destrudo set the space-time realities for the dynamic interactions among the three drives, setting the boundaries within each as libido operates in consciousness and unconsciousness. These two drives of construdo and destrudo circumscribe the possibilities of positive, pleasurable relationships; negative, unpleasant relationships; or none at all. These differentiations also indicate the difference between normal and pathological functioning in space-time. Libido's happy functioning is dependent on which of the two drives (construdo or destrudo) is predominating in the interpersonal situation, in the individual's psyche. One can facilitate great passion between two individuals (construdo); the other (destrudo) can facilitate rejection by both parties, and no libido will flow.

THE TIME FACTOR IN FREUD'S
UNCONSCIOUS DREAM WORK

Freud (1900/1953a) dealt with time repression in dream thoughts in *The Interpretation of Dreams*. He linked intermediate memory associations in a person's memory backward in time to earlier events in their childhood past, real and imaginary. These include construdo as well as destrudo fantasies and dream fantasies from childhood.

146

Such determining factors from construdo's past fantasies or past dreams are backwards in time. Dreams show a time order in which earlier mental stimulation from the memory system precedes and is more influential in dream formation than later stimulation from an individual's life. Freud (1900/1953a) said there was a backward pull or direction to the dream influence in time, which he called regression. Regression can be observed when a person's consciousness is dominated by past, earlier-in-time unconscious construdo fantasy determinations or dream determinations of what was reality.

Paradoxically, Freud's concept is that a dream represents the fulfillment of a wish. It indicates a forward unconscious temporal sequence of events leading to the dream. The wish has to first exist before its later sequential fulfillment can be expressed in the dream. So too, the daydream (a product of construdo) is forward in direction of its conscious fantasy, while it has a backward-in-time unconscious basis created by the daydreamer's unconscious destrudo, which gave rise to this forward-in-time conscious construdo. The daydreamer strives to undo, or break apart, destrudo self-perceptions of a negative, self-derogatory nature, that occurred in some present event and aroused negative memories. Even before drive relationships reach consciousness, there is already an interaction between the conscious and the unconscious aim directions of the drives.

Significantly, the drives (construdo and destrudo) move in opposite directions under these conditions of forward in consciousness and backward in unconsciousness. Thus we see both time moving backward in a dream and time moving forward.

Earlier, we stated that the operations of subatomic particles in the atoms of our brains define the operations of mental events within a dream. The space-time dimension allows us to find coordinates for dream and daydream events.

Besides the temporal determining factors of a dream, Freud spoke of the locality, or the spatial factor, of the dream scene or image. He did not, however, explain why one dream scene is evoked regarding an intermediate associational link versus another, either of which could be associated to the-earlier determining event of the dream thought.

Hawking (1988) said that "time is not completely separate from and independent of space, but is combined with it to form an entity called space-time" (p. 23). The space-time dimension applies to the reality of the universe beyond us, to us on earth, and to the subatomic operations within our brains, as these three determining realities result in the intermediate unconscious influences that confound individuals in their efforts at conscious direction of behavior. The human space reality in the universe, following relativity theory, indicates space is curved by the gravitational forces created by the celestial bodies of the universe (Hawking, 1988). Within an atom the spatial location of a subatomic particle is affected by its proximity to other subatomic particles.

Davies (1982), in *The Accidental Universe*, indicated that if a single subatomic particle is moving along a given path within an atom, prompted by a given quantum force, and it passes a second subatomic particle moving along another path because of a different quantum force, and it comes close to the first subatomic particle's path, then the first particle will be disturbed by coming close to the second particle. The first will aim a messenger particle at the second, which exchanges some quantum energy between Particle 1 and Particle 2, because of the disturbance aroused in the two particles' paths due to their proximity. Similarly, in the unconscious, when two mental fantasies are side by side, they will, because of construdo, combine or connect. If they are also similar, they will connect and construct a larger fantasy.

Although the position and course of one particle within an atom affects the position and course of a second particle (passing it in close proximity), when both particles are similar the grouping force or uniting influence is even more compelling. Theenergy that emits the messenger particle will disturb the path of the first particle. The energy transaction of the second particle requires it to absorb the messenger particle into itself, disturbing its original motion pathway (Davies, 1982).

Hawking (1988) described an experiment in which a single electron is directed at a screen with two slits in it. The single electron appears to pass through both openings at once, which can be explained in terms of wave particle duality theory. Hawking said that physicist Richard Feynman conceptualized the phenomenon like this: a particle is part of a wave of energy in which going from one place in space to another is seen as taking every possible pathway (p. 60). The logic or reality within an atom is different from the reality on our earth. The inference from physics is that space is not absolute within the atom.

The particles involved in the gravity force – the conjectured gravitons exchanged between bodies in the universe – affect each other similarly, as well as the space around each other, just as messenger particles exchanged between subatomic particles affect the course and pathway of particles within the atom. In either reality – in the universe or in the atom – space is not absolute or without external influences of surrounding bodies or particles within the atoms. Again, similarly, with respect to the personality feature of the unconscious, one thought affects other unconscious thoughts that are in close proximity. This grouping principle exemplifies how construdo operates in the unconscious.

Hawking (1988) noted that particles in the atom that cause binding (attracting) or repelling forces are particles that have no mass. This is apparently, according to TDT, the underlying subatomic basis of construdo's connecting and disconnecting in the unconscious. Conscious psychic operations are in turn affected. Hawking (1988) stated:

> In quantum mechanics, the forces or interactions between matter particles [dynamic interactions, between particles that have mass, or weight, or density]...such as an electron...emits a force-carrying particle. [Remember, force-carrying particles have no mass.] The recoil from this emission changes the velocity of the matter particle. [That is, when the matter particle sends off a force-carrying particle, it recoils like a fired rifle.] The force-carrying particle then collides with another matter particle and is absorbed. This collision changes the velocity of the second particle [with which it collides]. (p. 71)

One moving object colliding with another, changes the course of both. The resultant position and velocity of the changed particles in space-time create new coordinates of the collision event.

From these three influences from the forces of physics – from subatomic particle operations within the atoms of our brains, the realities of our planet, and the influence of universal forces on our unconscious – construdo creates a dream, moving from one level of the three influences to another, causing changing space-time in the dream. The regression of the time factor to significant events in an individual's life as intermediate associational scenes and images shifts the coordinates of the space-time dimension in a dream. The coordinates that created the dream can be determined. The forces on the dream formation cause it to stop intermittently at various places in space-time: a mental regression.

The uncertainty principle of modern nuclear physics explains that the position of a subatomic particle in space can be precisely determined when its speed is unknown. Its speed can be precisely determined when its position is unknown. Knowing one condition, either the speed or position of the particle in space, precludes knowing the other condition of the particle. The implication for the space-time dimension in dream formation – in the unconscious – is that to predict certain factors precisely precludes knowing the others precisely. So, knowing the time factors that create the dream, precludes knowing the determining space or event factors creating the dream, and vice versa. With the construdo drive, which affects the unconscious, time moving forward is an ego reality. In the unconscious, time also moves backward by destrudo influences. Libido joins in between the construdo's forward-in-time influence and destrudo's backward-in-time influence.

Dream interpretation in psychotherapy must involve sorting out the extent of these three influences of physics on construdo's creation of a dream. When the three influences come into play, when events recalled in an individual's life cause dream transformations, it tells the dream interpreter that the event had a powerful effect on the individual's life. Cumulatively, the effects may have led to a neurosis, a personality disorder, or a psychosis.

SPACE-TIME DIAGRAM

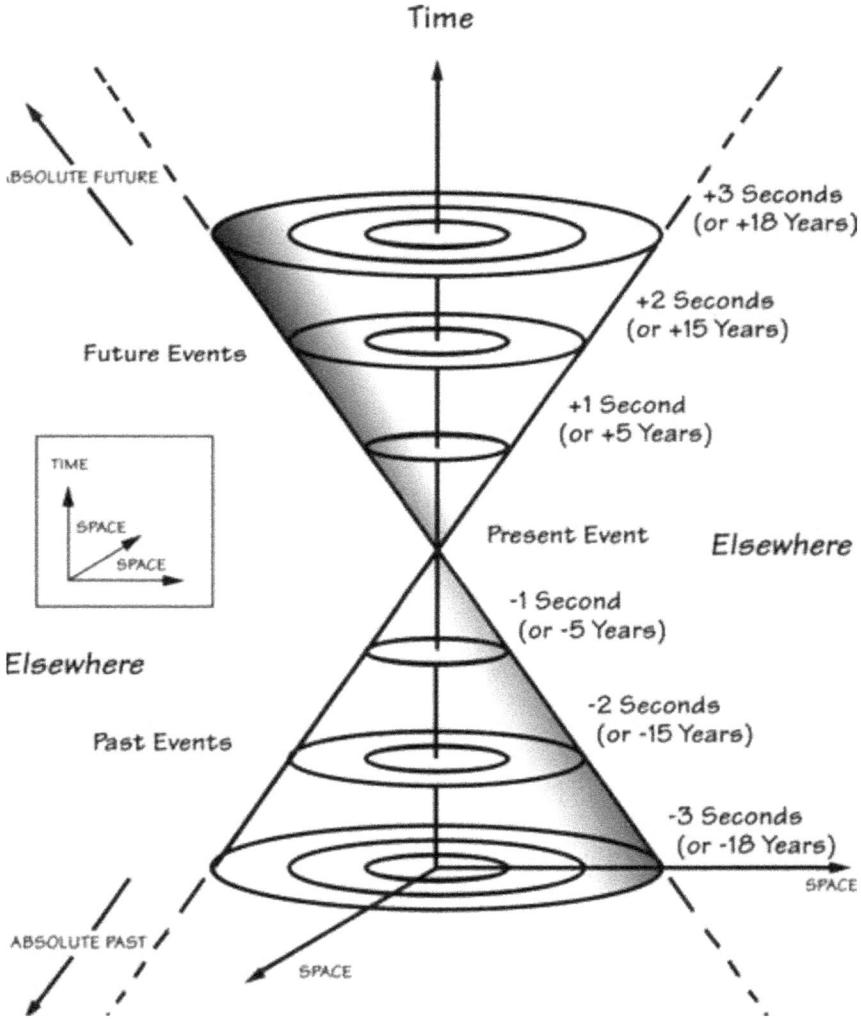

Figure 2. The space-time dimension in modern physics.
Reproduced with permission given by Kevin Rock (10/18/ 2015)

CHAPTER 13

BIOLOGICAL LIBIDO DRIVE CONTRASTED WITH DRIVES DERIVED FROM PHYSICS

INTRODUCTION

The interplay between the biological, procreative libido drive and the drives derived from physics – construdo and destrudo – reveal profound differences in the way they will affect us even billions of years from now. As noted throughout this volume, Hawking (1988) proclaims that when the universe reaches the point in its present expansion that it begins to collapse or contract back into itself, people will remember the future and not the past.

Because construdo is formed around future fantasies, the change to memories of the future may augment the strength of construdo. Thus people on our planet will show stronger construdo drives, which will presumably lead to their acting more constructively. For example, because they will not remember past events, they will be less angry. Destrudo, of course, will be weakened (less destructiveness) to the betterment of people because, of course, civilization progresses through constructiveness. In the meantime, while the universe is still expanding, humans want the ability to know and guess the future, which is implanted in our minds (our thinking) by the unconscious – by the fact of its forward-in-time existence (is an indication of unconscious potential and abilities). In the timeless unconscious, what is "now" and what "will be" (billions of years from now) is in close juxtaposition, because time is meaningless here.

That such a potential ability regarding the future might be revealed long before it exists in reality is understandable considering the past and future occurrence of an event as is revealed by phenomena in physics. Hawking (1988, pp. 24-32) explained this phenomenon: the past influences and directs what later occurs in reality in physics in terms of a present event (see Figure 2 at the end of Chapter 12). According to this sytem a present-day mental event defines a definite, prescribed future mental event. Thus the ultimate result of knowing the future will be a better understanding of the present mental event or reality.

Following influences from the realities of the past, as can be observed in the light cone (see Figure 2) of a present mental event as well as in the light of phenomanological future events, is the second set of realities that influence everyone and is the predictive outcome based upon the Hawking declaration. We know that factors from present realities influence an individual's conscious psyche. Therefore, both past and future influences on the unconscious psyche, as well as the present-day influences, must be taken into account in order to ultimately and fully understand the reality of a current mental event.

All these clarifications and distinctions describe human mental events in terms of the space-time dimension of physics. All normal or healthy mental events fall within this description and parameters of the space-time dimension. Construdo and destrudo changes occur along with normal biological or libidinal changes in the life cycle, as in the change from childhood to adolescence, from adolescence to young adulthood, when adults reach middle age, or when the middle aged enter older age. The fact of reaching each critical age stage prompts severe disruption in both the construdo drive's forward-in-time influence and the destrudo drive's backward-in-time influence on the ever biologically changing human being. These forward-in-time and backward-in-time influences happen whenever an anticipated future reality is imminent and therefore must be dealt with in terms of plans or fears of what might occur. Dealing with the future in this way during such periods of flux is normal or healthy, but can lead to temporary pathological behavior when the transition is in progress as a result of aroused fears. Also, psychoanalysis understands that transitions or transitional periods are usually difficult to manage psychologically. These periods also produce the sense of ambiguity that contributes ultimately to such aroused fears.

According to Triadic Drive Theory (TDT) these drive interactions of construdo and destrudo allow the influence from within our brain's atoms as well as the influences from the greater universe's influence on our unconscious to affect the person in such transitions because as stated, in the psyche everyone deals not only with present life issues but also past happenings leading up to the present and intended or hoped-for future occurrences.

In anxiety neurosis, someone considers an issue fearfully: "What if this or that happens?" This anticipation is governed by the mere possibility of something bad about to happen or by the probability that a present or future occurring event is or will be too much for the person to handle. In the case of hysterical neurosis, where perhaps a past erotic or destructive event was more than could be coped with, the person is likely to see a present event in terms of, or associated with past coordinates in the space-time dimension, so that feeling a similar inability to deal with the event that is taking place relates to this bad experience of that past event.

FEELINGS OF ALIENATION

One consequence of the duality of human drive origin (construdo and destrudo) is that the larger parts of the forces that drive us are not inherent to life. When people sense this duality, they may become aware of feelings of self-alienation. Having been created by forces of physics, construdo and destrudo operate in ways similar to forces of physics even in inanimate entities. Although our bodies and brains are biologically and genetically recreated, and these construdo/destrudo drives operate in what might be called a biological environment, they are not dominated by the forces of biology. They show stages of growth and development – a biological phenomenon – but at each stage the forces of physics reappear and are dominant. On the other hand, libido operates more in terms of living and growing things. Libido, while it has its own terms and principles of operations as described by Freud in his *Three Essays on Sexuality* (1905/1953b), is also pulled into dynamic interaction with construdo and destrudo by the construdo drive. Thus, libido falls under the influence of forces of physics as well as its own force of biology. The sensing in the human ego of such discrepancies between the two different mental energy sourcescreates, it is hypothesized, from such discrepancies, feelings of alienation within ourselves. This discrepancy between both mental energy sources and its relation to feelings of alienation occurs because the mental energy sources construdo and destrudo are so completely different from each other.

A CASE EXAMPLE

One adult patient in psychoanalytic psychotherapy for three years had with shame and guilt recounted his numerous affairs with women during his young adulthood and wondered what purpose in his life they served. During his therapy, he insisted that he was not allowed to express hostility openly, because of his mother's and stepfather's prohibitions and injunctions. Therefore, when his behavior as an adult provoked hostility by others toward him, he did not react with direct object-destrudo toward them – did not behave as though he wanted to destroy them.

One day, while recounting his philanderings as a young man, he mentioned that at times he was "a bad dude" with women. He explained that he belittled and degraded his girlfriends, calling them "stupid" or "dumb," (as his mother had called him when he was a child). Another example of his "bad dude" behavior could be seen when he'd have sex with one woman, then tell her she'd have to leave the apartment because another woman would soon be arriving. Psychoanalytically it is understood that his revenge toward his mother with these substitutes in his young adulthood, reflects his hostility toward women, generally.

As analysis progressed, he began to recall the feelings of triumph he experienced when he was in control of these triangular situations with women. Ultimately, he brought back memories of the primal scene. He had gone to his mother's bedroom when he was four years old and had seen his stepfather having sexual intercourse with his mother. His feeling at the time was that he had lost his mother. But in these later situations with women, the analysis is that it was he who had one woman only to displace her in favor of a second, as his mother had done to him in favor of his stepfather. The libidinal, Oedipal triangle problem had reemerged only now he was in control of the situation, giving him a feeling of triumph.

He later recalled an intense rivalry and antagonism toward his older sister, who, his mother said, had "more sense in her little finger than he had in all of him." His destrudo aim toward women had always been suppressed in his recollections except for his deception of them. He was more ashamed of these destrudo aims than by his libidinal, "pointless" philandering. In TDT language the manifest construdo influence here and the repressed destrudo influence in his unconscious formed the boundaries to his reality in which his libidinal Oedipal conflicts were reenacted. He emerged victorious in the situation over the absent, fantasized, construdo-pictured stepfather, the third shadowy mental figure in his Oedipal triangle.

Psychoanalytically speaking (especially in the shadow of Triadic Drive Theory (TDT)), sees this patient's feelings of shame and Oedipal guilt leading to his repression of the hostile object-destrudo drive aim. Such a hostile object destrudo aim motivated his behavior in young adulthood toward women (or its unconscious reversal) in overcoming the stepfather in the figure of the second woman. In analyzing the dynamics here we can see not only the libido drive operation between the construdo and destrudo limits but also the past times and places in which his pathology originated.

What were his repressing factors? Were they his view of himself from his superego, or his parents' prohibitions against expressions of aggression, or his guilt over his incestuous strivings for his mother, or his guilt over his sibling rivalry with his older sister? His uninhibited construdo drive was still free to construct relationships that could satisfy the unconscious striving of his destrudo and libido.

Thus, construdo was still active in this patient's consciousness in constructing these relationships. His genital libido aims were also in his consciousness. But his genital libido aim toward his mother from a former time and place was not conscious; that is, his striker-destrudo aims toward his mother and sister, displaced to other women in his life, were not in his consciousness. These striker-destrudo aims were unconsciously displaced from another time and place in his life.

Certainly, his striker-destrudo aim was in conflict with his construdo and libido aims. But was it the conflict in his unconscious that prevented the destrudo aim from reaching consciousness? Did the aims of libido and construdo that were aroused by these women result in repression of his later destrudo long-range intentions ultimately to debase the women, as Freud (1912/1957d) speculated in his paper, "On the universal tendency to debasement in the sphere of love"? Was it a component of his love? Was the patient showing unconscious sadism? Any of these explanations are possible.

TDT does not accept any of them as completely explaining the situation; instead, we ask: "Are there mechanisms and characteristics in the drives themselves, which preclude contradictory or antithetical unconscious drive aims from reaching consciousness? If all three object drive aims entered consciousness at the same time, would the patient have felt himself to be treacherous, a sadist, or psychotic?" TDT posits that when all three drive object aims are present in consciousness, as presumably in this case – the constructive stage-object aim of construdo, the striker stage-object aim of destrudo, and the genital-object aim of libido – then an individual feels emotionally overwhelmed by seeking to operate in such contradictory directions.

Remember that the three drives originate in the unconscious. Because they exist in the unconscious, arousing them pulls some of their energy into consciousness. When that happens, are there psychic mechanisms that limit and regulate how much of their energies can enter consciousness at the same time? What stages of each drive's object aims can enter consciousness together? Observation indicates the issue is how many unconscious aims can come into consciousness simultaneously. When all three drive aims entering a person's consciousness are self-aimed, he or she is psychotic, out of our space-time, and in a space-time that differs from the forces in the universe, on our planet, and within the brain's atoms. Such a person moves in many different and contradictory directions or operates in many different places at once.

In TDT understanding, subatomic particle energies in the atom that power the three drives offer an answer to why human drive energies aimed at certain targets operate in particular ways. How these particles are observed is relevant. All subatomic particles share a property called "spin." Of the two types of spins, one has 0s, 1s, or 2-integer spin, the other has a half spin. These particles have been observed to spin as they move (Hawking, 1988). Only the half-spin particles possess mass, and they make up the matter in the universe. Other spin particles produce forces that act on matter particles. These spin factors prompted Pauli's "exclusion principle": two similar particles cannot occupy the same space and have the same speed or velocity, precisely as prescribed by the uncertainty principle.

The implication for drive theory is that two drives cannot be aimed at the same object-target at the same space-time and speed. Thus, there is actually no fusion of drives, as previously held in psychoanalytic theory. Instead, drives sequentially fire at the same target, the idea originating from the unconscious basis of the drives. Their action is sequential, not simultaneous.

In TDT the interactions among the drives' energies coming out of the unconscious result in friction between the drives. It is exactly this friction between the psychic boundaries of the three drives that produces emotions or affects that human beings experience. An orientation in the drives toward a specific external object or environment space results in our moods.

When creative construdo-object aims and genital or phallic libidinal-object aims exist consciously toward someone and self-toppler or self-striker destrudo aims are conscious or unconscious, then a person feels overwhelming love and emotional attachment. Construdo and destrudo both play the major role in determining what type of emotional balance a person will experience. Libido fits within these boundaries. The more the drives are directed toward an object, the stronger will be a person's behavioral expressions toward the object. Unconscious energies produce these drives. In its simplest aim, construdo merely combines mental events or phenomena. This is because construdo is derived from the forces within the atoms of the early fetal brain as detailed in earlier sections of this volume.

FREUD'S CONCEPTS OF THE INSTINCTS OR DRIVE

In "The Instincts and Their Vicissitudes," Freud (1915/1957a) discussed the nature and characteristics of a drive. Consider how the introduction of the construdo drive changes the interpretation of Freud's data and how the introduction of specific stages of the aims of construdo and destrudo allows for different interpretations based on specificity of states, including the two uterine TDT's libidinal stages: "forming" and "contactual."

Finally, consider why these interpretations of Freud's data are less in accord with the dynamic concepts of modern physics, on which this volume is based. Freud depended on the scientific doctrine of the determinism of his time specifically with respect to unconscious influences, rather than on the more dynamic concepts of modern physics.

Surely Freud was aware of the dynamic concepts developing in physics at that time (first quarter of the 20th century) because he incorporated some of these concepts in his work on understanding of "instinct." Nonetheless he continued to rely on the concept of determinism in his evaluation of the drives. Nuclear physics was in its infancy, so its hypothetical psychic implications and corresponding consequences were not generally known until long after Freud's death in 1939.

Hawking (1988) described physicists of the beginning of the 19th century as arguing:

> ...that the universe was completely deterministic... If we knew the positions and speeds of the sun and the planets at one time, then we could use Newton's laws to calculate the state of the solar system at any other time... [It was assumed] that there were similar laws governing everything else, including human behavior. (p. 55)

Freud's concepts of the drives were clearly overly influenced by such determinism of 19th century science. In his paper on the instincts, Freud's (1915/1957a) view indicated that the drives seek ultimately to avoid excitation. He used *inertia*, a concept of Newton's laws in physics, to indicate a drive's striving. His concept of drive energy was that it sought to release the tension the drive had built up in an individual (Freud, 1920/1955a). In "The Economic Problem in Masochism," Freud (1924/1961b) conceived of the destructive drive as seeking a point in time where its energies would be expended and further stimulation or excitation was avoided.

Freud (1915/1957a) said that progression back to self-aims of the "instincts" can be seen as in the libido returning to its original, primary narcissistic stage beginnings in autoeroticism where the person's own body is the object of its libido. TDT, in constrast, elaborates the issue of self-aims of the instincts, but must add that all three drives (construdo, destrudo, and libido) originate in combining with the self in the self-combining construdo stage of the uterine unconscious. In the same paper, Freud stated that expenditure or release of energy of a drive never involves the "whole quota" of the instinctual impulse. That is also the view of Triadic Drive Theory insofar as Freud also held that the release of energy comes as "successive waves" like "erupting lava." In 1901, physicist Max Planck, in considering light-wave energy and other energy waves, asserted that energy was not released continually or arbitrarily, but in "packets" – or, in his word, "quanta." Thus, Freud's views and terminology are quite similar to Planck's concept from physics. In his 1915 paper, Freud asserted that the earlier direction and source of a drive (like the autoerotic direction of libido) will persist to some degree "side by side" with later object-directions.

The second part of the thesis in this section on quantum discharge of psychic energy waves is that these drives will not discharge from the unconscious into consciousness and/or into the external world unless there is an equal replenishment of energy back into the unconscious psyche.In this way, a balance is maintained between the leaving and reentering sources of unconscious psychic energy. This condition adheres to the physics principle of conservation of energy: "Energy (or its equivalent in mass) can be neither created nor destroyed" (Hawking, 1988, p. 184).

TDT's concept of psychic energy drive release incorporates the electromagnetic force with the directional flow of electric current. "Current" was the word Freud (1915/1957a) used to describe how a psychic drive can reverse or flow back into the self. Our further point from physics is that an electric current cannot flow unless it is part of a circuit loop from its origin and back to that origin again.

The drive application of this essential concept is that drives will not release their psychic energy into a situation unless the drive can reverse direction to replenish itself. Because of this prerequisite, reversibility, the psychic drives perpetuate their energy source in the unconscious.This is not accomplished in a single drive, although no single drive will discharge itself unless it can reverse direction simultaneously. Instead, the three drives (construdo, destrudo, and libido) are constantly involved in replenishing their original psychic energy. Because each drive originated from a self-combining, differentiated first stage of the drive in any person, the drive energy is replenished when psychic energy is returned to the original, undifferentiated, unconscious energy source. Hence, putting new psychic energy into a drive that has not been released can replenish the lost energy.

TDT sees that in the human brain, construdo and destrudo energies derived from forces of physics recapitulate energy returning the construdo and destrudo energies to thier original source, or paralleling the energy transformations and development in the history of the whole universe derived from the Big Bang. At the same time, boundaries set by construdo and destrudo (derived from physics) define how libido (derived from biology) will operate in individuals.

NEUROPHYSIOLOGICAL DATA ON NERVE TRANSMISSION

This section will detail the neural physiological underpinning of connecting and disconnecting mental drive energies. While TDT cannot prove a direct link or continuity between neural transmission in nerve fibers and the transmission of a drive's energies from the unconscious into consciousness, it can at least indicate that neural transmission along the neuron is accomplished by a bioelectric force. A neuron's response to a stimulus is to develop an action potential or nerve impulse along a nerve fiber to another nerve, gland, or muscle cell, and thus create stimulation by release of a chemical substance – a neurotransmitter – from the tips of the nerve cell fibers called telodendria (Hartenstein, 1976).

When a nerve cell (axon) is unstimulated, it has a negative charge (−) within the cell membrane covering the cell and a positive charge (+) on its outside. The membrane covers the cell's entire length. The positive–negative difference between the interior and exterior of the cell is important here. There is a higher concentration of sodium and chloride positive ions outside the membrane than inside. At the same time, the concentration of positively charged potassium ions and negatively charged proteins is greater inside the nerve cell membrane than it is outside. This condition is the "resting potential" of the axon. This potential is determined by the difference in positive ions outside the cell and the negative ions inside the cell along with the potential electrical energy interchange that can occur through the cell membrane. Exciting the nerve cell or firing it occurs when some of the negative ions from inside the cell are exchanged with some of the positive ions or charges outside the membrane. This exchange toward positive value or balance is called depolarization of the axon, the result of a nerve cell or fiber being excited and consequently discharging its energy.

If the electrical potential increases toward a negative value (or a balance), any response of the nerve cell is stopped, which inhibits stimulation or hyperpolarization of the cell membrane (Hartenstein, 1976). What follows is the propagation of bioelectricity in the dendrites of the nerve cell to its tips – the telodendria (Hartenstein, 1976). When a nerve cell is excited, the electrical interchange travels along the neuron; when inhibited, the electrical interchange dies out or blocks the firing of the axon or neuron, toward its target cells.

As stated, the above is the neural physiological underpinning of connecting and disconnecting mental drive energies. The disconnecting of construdo energies begins in subatomic particle operations of attraction versus repulsion, and continues to the neurophysiological level in nerve cells, as just described.

The idea of psychic drive energies as having its basis in physics follows. When a nerve fiber fires or discharges, this action potentially occurs when stimulation is intense enough in the space-time areas of excitation to pass the threshold for firing the nerve, and the cell reaches its action potential. Generating an action potential from the nerve cell's resting potential changes the electrical charge of the cell membrane from −70 mV reversing polarity to +30 mV – a change of 100 mV (Hartenstein, 1976).

The generated action potential discharging of the nerve cell, however, is not a flow of electric current but "the result of a chemical chain reaction in which sodium ions rush into the axon while potassium ions rush out" (Hartenstein, 1976, p. 81). Firing of the nerve cell occurs in successive bursts of energy lasting 2 to 3 ms (Hartenstein, 1976). It is not continuous, just as Planck (1901) asserted about the waves of light energy being discharged in quanta of energy, just as Freud asserted about drive energies operating in wave quota, and just as TDT has asserted about the discharge of psychic drive energy.

After each firing of the neuron (lasting 2 to 3 ms) there is a reversal of ionic flow wherein the:

> ...membrane pumps out sodium ions and pumps in potassium ions. This pumping action continues until the resting potential [of the nerve cell] is achieved again, also within milliseconds. Once the resting potential is nearly or entirely restored, the nerve cell can discharge again. (Hartenstein, 1976, p. 81)

Thus, there is a similarity here to what has been asserted about psychic drives' energies and the need for pathway circuits to return the flow of energy back to their origins, before they will again discharge or release energy. TDT calls it the replenishment principle.

The following is an example of a neurological underpinning of a construdo connecting and disconnecting energy circuit. When a nerve impulse reaches its target at the telodendria of a neuron and comes to a synapse or an infinitesimal gap (0.00000001 m) between it and a target cell (a muscle cell, gland cell, or another nerve cell), a neurotransmitter reacts with the membrane of the target cell. Again, a return flow of energy occurs. For example, if a nerve impulse stops at a synapse with a skeletal muscle target cell, then the chemical neurotransmitter acetylcholine is released, which diffuses across a fluid gap passing into the muscle cell membrane. The muscle cell membrane then becomes more permeable to an inflow of sodium ions and the outflow of potassium ions, depolarizing the muscle cell.

The level of acetylcholine released by the telodendria of the neuron determines the number of action potentials traveling along the neuron. The longer and more intensely the neuron is being stimulated, the greater its action potentials and the more acetylcholine will be released; thus the greater will be, for example, the depolarization of such a muscle cell membrane (Hartenstein, 1976). The muscle contracts according to the amount of acetylcholine released:

The membrane of the muscle cell cannot remain depolarized indefinitely or it could not be reset to generate another action potential to produce another contraction. Ordinarily, when acetylcholine arrives at the muscle membrane from the nerve endings, two things happen: (1) Acetylcholine reacts with the muscle membrane, making it more permeable to the inflow of sodium and the outflow of potassium. (2) An enzyme in the muscle membrane, called acetylcholinesterase, reacts with the acetylcholine and splits it into two parts, acetic acid and choline. The muscle membrane potential then returns to its original value and the muscle membrane is ready to fire again. In short, the effect of the acetylcholinesterase is to reset the membrane for another action potential. (Hartenstein, 1976, p. 82)

Again, the foregoing is a neurophysiological underpinning of a construdo connecting and disconnecting energy circuit. It shows why and how this occurs, neurophysiologically. But the psychological results on construdo have farther reaching results. These neurophysiological processes explain why neural transmission occurs in energy waves rather than continuously, just as TDT asserted drive energies are transmitted, and as Planck (1901) asserted that energies of physics were transmitted. In this instance, the effect of acetylcholinesterase causes resetting of the target cell (a muscle membrane) to receive another action potential charge from the nerve cell.

Further support for TDT's notion of a drive flow needing a means for replenishing is further found in the neurophysiology of neural transmission:

The second way in which a neurotransmitter may be inactivated is by being taken back into the neuron from which it was released. In the central nervous system, for example, some of the noradrenaline which is used at a synapse is inactivated by enzymatic modification. Much of it, however, is inactivated by the process of reuptake. Presumably it is energetically cheaper to bring the molecules back to the presynaptic molecules. Inactivation by reuptake may be an important conservation feature, especially in the brain, where millions of synapses are occurring every millisecond. (Hartenstein, 1976, p. 83)

This energy conservation confirms what has been asserted about the conservation of psychic energy, derived from the physics principle of conservation of energy. It applies after the drive energy has come out of the unconscious psyche.

To extend this discussion of neurophysiology, consider a situation where a nerve impulse is inhibited at a synapse, as in the parasympathetic nerve system innervating the heart muscle. Again, acetylcholine is released when the nerve impulse or action potential reaches the nerve cell telodendria. But in the heart there is a different result than a skeletal muscle synaptic junction. The cell membranes of the heart muscle hyperpolarize and become inhibited (Hartenstein, 1976, p. 84). Hartenstein (1976) said the reason for the difference is not known; only "that it exists." He asked:

> Why are there two kinds of reactions to a stimulus, excitatory and inhibitory? The answer is that many target cells, such as heart muscle fibers... and countless nerve cells in the central nervous system are innervated by two or more nerve cells. One nerve cell makes an excitatory synaptic connection with the target cell. The other forms an inhibitory connection. In this way, the electrical activity of a target cell can be stepped up or down as needed. (Hartenstein, 1976, p. 84)

This neurophysiological finding of two types of nerve cells (one leading to excitation, the other to inhibition of neural electrical activity) parallels TDT's concept of a drive discharging itself and subsequently replenishing itself. It is also consistent with construdo's connecting and disconnecting, reminiscent of Freud's concepts of cathexis and anticathexis (Freud, 1900/1953a, p. 605). It also follows Freud's (1933/1964) concept of an instinct or drives being "inhibited in their aim" (p. 97).

RETURNING TO TRIADIC DRIVE THEORY

It has been asserted that it is in the nature of the three drives (construdo, destrudo, libido) to be linked with reality by the construdo drive, which is constantly, dynamically interacting in the unconscious to reenergize these drives. This interaction perpetuates their energy, paralleling the biological libido's means of perpetuating humans through reproduction. However, when one, two, or three of the drives are aroused from the unconscious and enter one's conscious functioning, the uninvolved drive is understood to be self-reenergized by mental energy returning to the unconscious source of the drives.

Drive energies, while not focused on in their response to stimulations, are nonetheless aligned by the replenishment principle in the interaction with drives not central to the conscious drive response. This psychic energy returns from the external world through the drives. It re-equalizes the energy of the responding expended drives, because the unconscious sources of the three drives are connected by unconscious self-construdo, and they flow back into each other. Thus, psychic energy expended in consciousness will equal the energy replenished back into the unconscious undifferentiated drive energy origins.

TDT posits that there is an attracting to things in the arousal of connecting construdo and also a repelling from things in the arousal of disconnecting construdo. The attracting and repelling influences here are derived from similar processes in the operations of subatomic particles in the brain. These influences cause the drives derived from physics and underlie their operations in the unconscious. Remember, the influence of destrudo breaks apart connections of construdo in consciousness.

Assigning a positive attraction or a plus (+) sign to a drive connection, and a minus (−) sign to the negative disconnection of a drive, allows the inference that the human drives operate with an influence that is similar to the way the electromagnetic force in subatomic particles in the brain operates.

Compare the aims and stages of construdo and destrudo, the self-connected construdo impulses versus object-connected construdo. Dreams demonstrate transformations of these two drives in terms of space displacements and time displacements, and space-time condensations as is in the nature of the unconscious. Such examples involve transforming construdo and destrudo aims into different aims altogether. That is distortions of construdo and its combining, connecting links and transformations of destrudo aims to break apart, strike down, or topple an existing construdo link.

Why do transformations of intermediate thinking, concepts, and the ultimate aims of construdo and destrudo occur only in the unconscious? Freud (1900/1953a) outlined in *The Interpretation of Dreams* that the unconscious has its own principles of operation. We reassert that these principles are derived from the operations of subatomic particles within the atoms of the human brain, creating our unconscious − and from nowhere else.

Moreover, the answer evolves from the conundrum of why several psychic drive energies cannot all be discharged at once, at a single target or space, simultaneously. The question is why not?

The answer is found in Wolfgang Pauli's "exclusion principle" regarding particles within the atom: "Two similar particles cannot exist in the same state, that is, they cannot have both the same position and condition, and velocity, within the limits given by the uncertainty principle" (Hawking, 1988, p. 62). This means that the unconscious psychic energies of the drives cannot have the same target or aim in space that strive for or seek a target at the same time and with the same velocity. In psychoanalytic terms, it means the drives do not show fusion but rather sequential firing.

Applied to triadic theory, this means that two or three drives cannot discharge to the same arousal place at the same speed, which accounts for how the drives remain separated during their passage from the unconscious psyche into the conscious psyche. The drives do not fuse. Freudian psychoanalytical theory believed they did.

Because they cannot travel at the same speed or velocity when more than one drive is aroused, we can observe their expression in a time sequence, one after another. Drive ambivalence can be expressed when at one moment an individual loves the object (creative construdo) and the next moment expresses smasher-destrudo hate for the object. Neurophysiology helps explain this confusion: Some neurons are responsive to excitation, others to inhibition. The two drives, construdo and destrudo, lead to love in a creative-construdo advanced stage and to smasher-destrudo levels of hate.

Next, we will consider how Freud's (1915/1957a) views (in his "instincts paper") that deal with love, hate and indifference, sadism, masochism, scopophilia, and exhibitionism – amplified the characteristics of the drives we are positing. That is, the nature of libido functioning has relevance to construdo and destrudo influences.

FREUD'S OBSERVATION OF THE VICISSITUDES
AND OUTCOMES OF DRIVES

The first two vicissitudes of a drive, Freud (1915/1957a) observed, were that a drive can show "reversal into its opposite" and that it can "turn round upon the subject's own self" (p. 126). Repression, sublimation, and reaction formations against the drives are dealt with in other Freud's papers.

Demonstrating drive reversal to its opposite or turning the drive onto itself is taken up by Freud in terms of a change from active to passive, as in the change from scopophilia to exhibitionism. He said this began with an individual's looking at his or her own sexual organ, the penis or vulva/vagina. That simple act, he said, can lead to a reversal into scopophilia or voyeurism. That point illustrates the difference between TDT and Freud's conceptualization of the self stage of libido. Freud distinguished it as primary narcissism; TDT considers it a self-oriented stage wherein libido by self-construdo produces a libidinally charged reaction to self-libido. There's an alternating influence here between self-construdo and object-construdo in the unconscious (from looking at oneself to conscious voyeurism), or looking at others (libido connected to construdo).

In the reversal of sadism into masochism, Freud found masochism to be sadism turned against oneself. TDT asserts that self-construdo is coupled with pincer, toppler, striker, or smasher self-construdo. The self is the object. Libido may not be involved, or may be brought into the drive constellation by construdo.

Freud observed that masochists find pleasure in assaulting and torturing themselves. Self-construdo links self-destrudo and self-libido. The object of these torture fantasies is changed from someone else as object to the self (i.e., self-destrudo).

What is behind the constellation of the three drives that produces a self-conviction of masochism in behavior? Remember, the self-combining construdo stage returns to the fantasy of being the object of destrudo torture and control. In Triadic Drive Theory, that's because in utero construdo first combined with the self and cannot sustain the separate development to connect with objects in the external world. The drive energies slip back to the uterine self-stages of the drives, as when the self was the earlier object aim. Remember that in triadic theory all three drives begin as self-directed in utero. That is, the self is the first object of the drives' aims. Self-combining and self-connecting stages of construdo set the stage. Destrudo runs its course against the self object. When libido gets involved, the behavior leads to the discovery that there can be pleasing torture in masochism. Masochism is the product of self-destrudo, self-construdo, and self-libido. Only then do we observe anything that looks like what is described as masochism.

The change Freud observed, from active to passive, is a change back to the earlier self-construdo stage, because the drive cannot maintain its forward development. The drive regresses to a position it can hold, which in terms of energy output is passive. The drive certainly is not as active as it would be if construdo and destrudo were simultaneously, and energetically, advancing forward in time toward the same object.

The reversal of a drive found by Freud (1915/1957a) involved a change from an object aim in consciousness to an opposite aim in the unconscious. In triadic theory, when drive energy returns from consciousness back into the unconscious, its aim can be reversed by the unconscious. These are the changes in energies that are created in the unconscious.

Regression to the self-combining stage of unconscious construdo with its reversal tendencies can cause a turning point in the drives, whether it's combined with libido, destrudo, or construdo itself. This leads as Freud noted, to the conscious observation of a drive becoming its opposite.

Another source of the turning of a drive comes from the destrudo's backward influence in space-time. It causes an unconscious space-time influence that is opposed to construdo aims. The influence is reflected in the different drives turning into their conscious opposites.

TDT's conceptualization of the self-combining stage of construdo and its combining with destrudo and libido does not require as many transformations of psychic events, nor does it require the entirely different or special condition such as the state of narcissism, as Freud (1914/1957c) proposed.

Freud's pleasure-unpleasure principle, however, is essential to the conceptualization of the three drives and their operations in terms of the drives' conscious attracting and repelling influences. When these unconscious and conscious influences are in tandem or congruent, they show the strongest energy. It is an instance of the uncertainty principle in action.

At the time of Freud's writings, there was no concept of the two drives derived from physics, versus the biological drive libido, nor of a construdo drive. Nevertheless, differences in the drive's operations as seen in TDT are also seen in Freud (1915/1957a). He saw drive reversals. The perception of such reversals led TDT to conceive that these changing drive energies always go back in time to the undermost energy origins of the drive involved. This principle of return to undermost energy origins helps explain the vicissitudes of psychic drive expenditures of energy and the final direction such drive energies follow. Therefore more is involved than the "repetition compulsion" to repeat or return to earlier, hurtful, unpleasurable states requiring corrective change.

The psychic drives reflect the tendency of forces in the universe to return to what first caused or created them. The drives are carried on the wave of this energy-direction to the drives energy in the human brain. We also find here the uncertainty principle in operation.

Such a nostalgic impulse "to return to origins" supersedes how the drives operate based on psychic factors derived from physics, underscoring the more extensive influence of the universal forces of physics on the psychic drives and their outcomes. TDT awaits future confirmation from physics. This tendency has not specifically been observed by physicists, other than by inference from Hawking's (1988) concept of the universe reversing from Big Bang to Big Crunch, his work on stars that collapse back into black holes, and his view that there is evidence of a large black region near the center of the Milky Way galaxy. All of Hawking's (1988) observations and conclusions rely on his proposition that the universe has no specific boundaries. TDT infers that as one progresses backward from universe to galaxies to stars and to smaller units in the vastness of the universe, all the way to human beings, the urge to return to earlier undermost energy origins can be observed.

Remember, the construdo drive reinfluences the self from the earlier original stage of each drive's beginning at its own uterine origins. In accord with the second law of thermodynamics, in a closed system, a drive's influence moves from construdo (which orders things into combinations, connections, and constructions) to destrudo, which ultimately breaks apart these combinations into disorder and disarray. It is the physics principle of entropy: Things go from order to disorder. People worry about "things not going smoothly" as they have planned, things that deteriorate with the passage of time. Hawking (1988, p. 145) called this progression and direction the "psychological arrow of time," the sense of time humans hold in conscious reality: we know what has happened, but we do not have a sense of what *will* happen.

CHAPTER 14

PARAMETERS OF DREAMS AND REALITY

THE DREAM PARAMETERS

Freud (1900/1953a) believed that dreams could be reduced to erotic dream thoughts. TDT adds the concept that within a single dream, or group of dreams covering the same idea in a short time period, the construdo and destrudo drives define the limits of the space within which a dream occurs. Human existence in reality, with respect to construdo and destrudo drives, shows similar boundaries.

A Female Patient's Dream during Psychoanalytic Therapy

A patient remembered a dream in which she was watching a television program that interested her. Her father came into the room and changed the channel, saying he didn't want her to look at the program. She couldn't understand his action and resented it, but she said nothing.

Her associations with and reactions to the dream indicated that her father's authority over her life was generally arbitrary, of an "old world" character. He tended to treat her like a little girl, although she was in her late 20s. He wanted her to remain in his house until she married. However, he tried to supervise her relations with young men and usually made these suitors feel uncomfortable when they visited her. He also disparaged them in various ways, until she wondered if she could ever have a satisfactory relationship with a man while especially she was living at home.

Nevertheless, she was in conflict about her motives for moving out of her father's home versus her motives to please her father, to do as he wanted, and to be "a good daughter." He told her he would disown her if she moved out. This frightened her. On the one hand, she felt she should move out into the adult world, lead an adult life, and establish herself as an adult. But she felt guilty about opposing her father. She was not destrudo-driven to fight for herself.

The boundaries of this young woman's life lay between wanting to join the world of other young adults as an independent woman, having free contact with men, and being in a construdo-constructed relationship with her father and family (the construdo boundaries, secondarily supported by libido), versus the need to oppose her father in an unexpressed destrudo stand against him (the destrudo boundary). Previously, she had reported that her father actually talked as if he was jealous of her attention to her

171

suitors. He talked of losing her attention and correspondingly she had no conscious awareness of her father's apparent unconscious participation in his parental Oedipal relationship with her.

In discussing the dream with her, I suggested that trying to learn something about the world by watching the TV program reflected her drive to join the world around her, a derivative of the construdo drive reflecting her desire to connect with her surrounding reality. But her father was blocking this drive-aim. When asked if she thought her father opposed her desire to learn about the world, she was hesitant about agreeing or admitting it.

I had noted her silence in the dream, expressing no resentment at her father for willfully changing the channel she was watching. She responded by indicating her unwillingness to express her resentment or to act aggressively. By not acting, she was implicitly declaring that her destrudo expression was blocked. Any explicit response equaled parental disrespect, which she was loath to express.

In the session when she presented the dream, we discussed the place of construdo, destrudo, and libido in terms of her Oedipal problem. Yet shortly after this session, she rejected further psychotherapy. I wondered at the time whether confronting patients with the motives of all three drives at once was too much for them. That might equal leaving patients with no drive within which to conceal themselves—an "uncovering" therapy that moved too fast, overwhelming patients with their own pathology without leaving them a place to hide. The idea has not been resolved.

However, in this case, I had an unusual opportunity to discuss the patient with a colleague whom the patient subsequently consulted. His immediate impression was that she wanted to be assured that he would not be too analytical in pressing his inquiry into her condition. My colleague's impressions confirmed that I had overwhelmed her in presenting the status of her three drives, challenging both the conscious and unconscious levels. Thus, premature inquiry had apparently triggered factors in the patient's psyche too powerful to be dealt with in one session.

It seems likely that her dream and her reactions to my dream interpretations reflected the reality of her life, showing how space and time anchorings of the conflicts with construdo and destrudo in her life circumscribed the boundaries within which she acted out her life's dramas, instead of living life spontaneously. It is the construdo and destrudo dream thoughts that define these limits. In psychotherapy, then, this patient was asked to consider factors she herself did not want to consider, factors that were transferentially tantamount to her father telling her to do things that she knew were not in her best interest.

How people deal with and react to their construdo and destrudo drives defines the limits within which they deal with their psychological problems. Inside these boundaries, their dreams reveal problems they are both unconsciously and consciously ready to confront. Even so, as this young woman's dream indicates, that does not mean these patients are ready to confront the constructive and destructive drive limits of their unconscious dream thoughts. These confrontations may be beyond a patient's unconscious tolerance. Thus, the limits should be tested during therapy but never exceeded.

The dream boundaries of construdo and destrudo suggest the limits within which a person will consciously participate in his or her surrounding reality. To extend this concept, is to consider that how people participate in society reflects these drive limits at any given space-time; that is what people will join and what they will not join within their construdo-drive and the destrudo-drive limits.

There is a similarity here to Hawking's (1988) concept of a universe with expanding boundaries: the three psychic drives have no definite, established boundaries but are yet limited by how libido will operate between the indefinite boundaries of construdo and destrudo. This demonstrates the parallel between how the forces of the universe operate and interact, and hypothetically, how universal forces may affect the human psyche. The concept parallel's Freud's (1907/1959c) view that what is observed in the society of human beings will also have occurred within an individual's own psychic development.

Freud's Irma Dream

The first dream Freud (1900/1957a) presented in *The Interpretation of Dreams*, wherein he reports his own dreams, was his "Irma dream."

A large hall; numerous guests, whom we were receiving. Among them was Irma. I at once took her on one side, as though to answer her letter and to reproach her for not having accepted my "solution" yet. I said to her: "If you still get pains, it's really only your fault." She replied: "If you only knew what pains I've got now in my throat and stomach and abdomen – it's choking me" – I was alarmed and looked at her. She looked pale and puffy. I thought to myself that after all I must be missing some organic trouble. I took her to the window and looked down her throat, and she showed signs of recalcitrance, like women with artificial dentures. I thought to myself that there was really no need for her to do that. She then opened her mouth properly and on the right I found a big white patch; at another place I saw extensive whitish grey scabs upon some remarkable curly structures which were evidently modeled on the turbinal bones of the nose. I at once called in Dr. M., and he repeated the examination and confirmed it. ... Dr. M. looked quite different from usual; he was very pale, he walked with a limp and his chin was clean-shaven. ... My friend Otto was now standing beside her as well and my friend Leopold was percussing her through her bodice and saying: "She has a dull area low down on the left." He also indicated that a portion of the skin on the left shoulder was infiltrated. (I notice this, just as he did, in spite of her dress.). ...M. said: "There's no doubt it's an infection, but no matter; dysentery will supervene and the toxin will be eliminated." ... We were directly aware, too, of the origin of the infection. Not long before, when she was feeling unwell, my friend Otto had given her an injection of a preparation of propyl, propyls ... propionic acid ... trimethylamine (and I saw before me the formula for this printed in heavy type). ... Injections of that sort ought not to be made so thoughtlessly. ... And probably the syringe had not been clean. (Freud, 1900/1953a, p. 107)

Evidence of Construdo and Destrudo Boundaries in Freud's Irma Dream Analysis

In his analysis of his Irma dream, Freud (1900/1953a) discussed the first scene of receiving guests, including Irma, on the anticipated occasion of his wife's birthday. This is a transformed construdo dream thought of wanting to join and combine with other people. As the first dream thought of the first scene, it defines one boundary of the dream. Such a boundary does not always appear as the first scene, because time and space might have occurred at any place in the actual or latent dream content.

Freud (1900/1953a) then explained his annoyed reaction to Irma, a destrudo thought, telling her it's her fault if she does not get well. Subsequently, he discussed examining Irma's throat by the window in the dream. His associations lead him to a throat examination he actually did make of a beautiful young governess, a woman friend of Irma's whom he held in high regard. He had once visited this woman and found her by a window, just as Irma was in the dream. The governess was a patient of Dr. Leopold M. Associations that combined dream material with reality and other aspects of Irma's case were prompted by the construdo-combining stage aim.

174

Other dream images led Freud to associations that recalled his early experiments with cocaine and recommending it to a dear friend a few years earlier. Misuse of the drug, Freud said, hastened the death of his friend and brought serious allegations toward Freud. His connection with these guilt feelings further emphasized his guilt about Irma's treatment. Further associations led him to comment, "It seemed as if I had been collecting all the occasions which I could bring up against myself as evidence of lack of medical conscientiousness" (Freud, 1900/1953a, p. 112). This "collecting" is a self-construdo indication, as well as a self-destrudo indication in the unconscious.

Freud's associations led him to realize he had reasons for feeling "ill-humor" toward his friends Otto and Leopold for rejecting an idea he had recently put to them. These thoughts are simultaneous construdo and destrudo dream thought fusions: the friendships reflect object-construdo, and the ill-humor reflects object-destrudo.

Dream thoughts stemming from construdo and destrudo vacillations continue and can be noted throughout the Irma dream narration.

Freud said he saw the chemical formula for trimethylamine in his dream. In his thought connections with this substance, he connects his friend, Wilhelm Fleiss, and his thoughts that Fleiss, "believed that one of the products of sexual metabolism was trimethylamine. Thus this substance led me to sexuality" (Freud, 1900/1953a, p. 116).

"Trimethylamine was an allusion not only to the immensely powerful factor of sexuality, but also to a person whose agreement I recalled with satisfaction whenever I felt isolated in my opinions" (Freud, 1900/1953a, p. 117). Freud's thoughts may also have been that the young widow Irma would be more amenable to his analysis if sexuality came back into her life, which might lessen his destrudo toward her and hers toward him. Freud is also feeling connected by his construdo with Fliess (i.e., in his effort to combine himself withfellow believers of an erotic linkage in dreams) and so he was disconnecting from Irma's disbelief of his analysis.

Believers were Freud's positive construdo links, disbelievers his destrudo links, breaking apart from these others in his environment. These two drives defined the limits of Freud's personal world. The Irma dream demonstrates vacillations between the construdo connections that joined Freud to his world and the destrudo/construdo connections breaking him apart from it. The combined connections retain the construdo drive power to disconnect itself, but construdo can also reconnect with another ideation. Reconnections in this unconscious sphere may reveal little in the way of logic or rationality if the connections remain in the unconscious.

The destrudo trend can be seen in Freud's analysis of the Irma dream when he commented, "Otto had in fact annoyed me by his remarks about Irma's incomplete cure" (Freud, 1900/1953a, p. 118). More direct statements of Freud's destrudo can be found in statements such as, "I revenge myself on Otto for being too hasty in taking sides against me... But Otto was not the only person to suffer from the vials of my wrath. I took revenge as well on my disobedient patient" (Freud, 1900/1953a, p. 119). Freud wanted to make Dr. Leopold M. look ridiculous. Freud is clearly referring to the by-products here of his object-toppler destrudo and its conflict with his constructing-construdo links with those supporters around him.

This is precisely the type of conflict Freud found in all neurotic reactions. Either construdo or destrudo at the same time might be connected with libido. The construdo and destrudo limits define here the world Freud was consciously participating in at this given space-time in his life. In addition, between construdo and destrudo boundaries, libido-aim can also be identified in what has been earlier described as "displacement upwards"; that is "looking down her throat" essentially meaning perhaps "looking up her vagina." This can mean that with construdo and destrudo boundaries, libido aims may also be operating.

Unconscious Conflict and the Atom

According to TDT inherent basic conflicts are found in the unconscious underpinning in the operation of subatomic particles—from annihilation of particles to their creation within the brain's atoms. The oppositional character of construdo and destrudo in the unconscious also creates the conflicting feelings we experience between constructive motives to hold ourselves and objects together versus concession to the destructive motives to break them apart within ourselves as well as with objects. This desire begins in utero. In our lives we often build things up, then break them apart. It is seen in a child of three or four building a tower of blocks and then, having succeeded, knocking them down with a sweep of the hand.

Anna Freud (1972) said psychoanalysis has never quite understood why children carefully build block towers of three or four blocks, then knock them down. TDT posits that the action demonstrates the unconscious conflict underlying human behavior and is reflected in the actions of human societies – namely, what humanity creates, it often tears down.

Construdo and Destrudo Parameters in the Two Dreams

In the two dreams presented above, my patient's dream and Freud's Irma dream, it can be seen that the conflict in the dreams is between joining the world around the dreamer constructively and destrudo's striving to disunite and break the bonds with figures that do not lead to productive, satisfying, and fulfilling relations. There is a construdo-destrudo conflict in my patient's dream – between wanting to learn about the world and join it in watching a certain TV show and her father's turning it off, and thus letting her see nothing of what she wants to learn. Her unconscious Oedipal problems fall in the middle of this conflict. She wishes her father would promote both her knowledge of the world and also her growth toward independence from him. Both are self-construdo wishes.

Freud's Irma dream, as he said, shows his fulfilled wish not to be held responsible as a doctor for her lack of a full cure. This wish is revealed through his annoyance, a destrudo impulse, with his friend Otto, who suggested that Irma was only partially better. Freud's reporting of his being friendly with Irma's family, the first scene in the dream, of he and his wife receiving guests, his comment that in the dream he turns against those with whom he is friendly (namely, Otto and Leopold), all indicate his construdo attachments, positive connections that change into destrudo connections with these figures. These drives set the boundaries within which life with them will be lived out, in the dream.

That Freud became annoyed with Irma, Otto, and Leopold in his "Irma dream" indicates the conflict creeping into the dream, of his not wanting to be accountable for the results of Irma's treatment. He is fending off implications in his dream that he will feel toward these people in the future.

Any erotic impulse toward the young governess, Irma's friend, is not analyzed, as Freud says he does not wish to "penetrate" such matters further – a very telling and suggestive choice of words in terms of their sexual connotations. But if such impulses were reported, they would take place between the extremes of Freud's conflict about Irma's treatment. Such parameters in the Irma dream stem from construdo and destrudo being in conflict with each other in the unconscious psyche creating the dream. In addition, of course, as indicated, libido aims can also be deduced in the displacement.

Such conflicts stem from the fact that in utero, while the unconscious psyche is forming and maintaining life, the drive union between libido and construdo to hold the forming fetus together; the destrudo drive ultimately strives to break apart such a union and its implications for the continued life of the organism. Again, according to TDT within the fetal cells of the organism, the potential influence of the atomic forces to break apart, and the influence of holding them together,formthe conflicting influences we have previously seen in subatomic particles.

The operations of subatomic particles within the atom indicate particles smashing into each other and annihilating each other in smasher-object destrudo fashion, then creating new particles in construdo-connecting and constructing fashion. TDT proposes that from within the cells of the brain and body there is a registering of these forces; the unconscious psyche responds to them. In the unconscious psyche construdo and destrudo come into existence as unified, organized, opposing psychic drive-forces.

REALITY PARAMETERS
OF CONSTRUDO AND DESTRUDO

The conflicts in Freud from an individual's unconscious follow from the opposition of construdo and destrudo in the unconscious. Much of an individual's life is directed and governed by happenings the individual supports, wants to follow, join with, become a member of, continue, participate in, and feel pleasure in so doing. We may imitate and follow the behaviors of others in buying certain homes, dressing in certain fashions, listening to certain music. We identify with a group or class and do what they do. All these behaviors are derivatives of construdo.

In contrast, there are parameters of destrudo derivatives from the reality that cause people to be firmly against certain life styles or a prevailing viewpoint to which they do not wish to conform. This is a boundary people draw in terms of what they do not want to do with their lives. Destrudo plus construdo combining in the unconscious create these destrudo limits. It can be tantamount to living one's life according to what one is against. People may be against a political party in power, groups of people at their job, an ethnic minority or racial group they dislike, or family members they dislike and wish to change. These are destrudo boundaries, but the energy behind such boundaries comes from the unconscious, and by definition people do not realize it.

People can be connected via construdo to hate groups and further their destrudo needs. They may strive to outdo a competitor in business, a political opponent, a rival in church, a colleague in their profession, or a mate in marriage. News reporters might report the flaws and shortcomings of a popular president. All the foregoing are destrudo derivatives of the striker and toppler-object destrudo stage-aims. Therefore destrudo can have far-reaching influences on people that are not obvious or easily understood.

Reality Parameters

Human reality limits as well as humans' participation in their own life are governed and directed by unconscious construdo and destrudo. The derivatives' results of these drives in humans' participation in the reality of their lives is more extensive than previously believed. It's far more than could ever be inferred from the influence of libido, as Freud believed.

That the unconscious plays a greater role in people's lives than they realize was one point Freud (1900/1953a) made in *The Interpretation of Dreams*. The point is repeated by TDT's introduction of how the forces of physics definitively influence humankind's unconscious as well as their subsequent conscious lives.

In summary, TDT believes psychoanalysis founded on Freud's drive theory is not intellectually consistent with findings of modern science in the absence of the findings of nuclear physics. In Triadic Drive Theory, the biological drive theory of Freud is updated by forces of physics. This creates two new drives, destrudo and construdo. In this sense, TDT seems to signal is a significant progression in drive theory.

CHAPTER 15

PSYCHOPATHOLOGY IN THE DRIVES

IMBALANCES OF FORCES FROM OUR UNIVERSE, OUR PLANET AND **OUR BRAINS' ATOMS** CAUSE DRIVE PATHOLOGY

The closer a drive response corresponds to the conscious psyche's reality, the healthier or more "normal" a person's response will be. The fewer distortions of reality to which the response is subjected the healthier a person's response will be. Psychopathology in and among the drives can be defined by the extent that our unconscious causes our drives to respond to other realities than those of our planet. Our planet's realities are our lives' realities.

Influences from the other realities in the unconscious after birth make it impossible for a newborn to accurately combine with many aspects of its surrounding world's realities, because of the constant shifting influences from these realities when perceived through their unconscious eyes. Only as the other two realities differing from our planet's realities gradually recede and separate in the unconscious during the first six months of life can construdo attach to a stable world of events and things with which it can consistently combine. This construdo event marks the beginning of Freud's (1911/1958b) reality principle and the origination of the newborn's conscious psyche, which develops out of what Piaget (1954) described as the "undifferentiated chaos" of the infant's initial perceptions.

With respect to Triadic Drive Theory (TDT), the influences from the two other realities – from the universe and from within the brain's atoms – remain in the unconscious, because they would render reality incomprehensible for the infant. The development of construdo leads the infant to combine more readily and accurately in reality terms with its surroundings, which reflects the construdo drive's basic aim.

Influences from the universe stored in the unconscious recall universal existence moving away from its origins in the Big Bang, what Hawking (1988) calls the "cosmological carrot of time" (p. 145). In parallel fashion in the unconscious drive energy moves outward, away from its origins of energy and psychic power into the three differentiating drives. This direction of human drive energy is comparable to the way energies operate in the universe. There is a parallel between the moves away from the energy source of the unconscious that with respect to TDT understanding follows the energy paths in space-time of our expanding

181

universe. In other words, these energies (unconscious and universe) expand, combine, and connect with new realities, just as the expanding universe creates new regions within itself.

It can be conceived that the unconscious registers that the forces of expansion weaken as the universe continuously expands and as the gravity pull on matter with the universe increases. Similarly, it can be conceived that humans experience these separation reactions when parted from familiar territory, loved ones, or any thing, person, or place with which they had felt mentally connected. In terms of TDT this sort of separation reaction may be considered as an anxious construdo reaction.

Space and time, while not absolute, are interrelated, affecting each other, thus creating the space-time dimension in our mind's reality. This interrelation of space and time translates into a perceptually shifting existence where nothing is as finite as the human perceptual apparatus wants us to believe. Therefore, space-time events are always uncertain and cannot always be predicted precisely. Unless these sensations are completely repressed by construdo (which determines whether it should combine and connect with what is being perceived), the very uncertainty of space-time events will create anxiety in a person's construdo reaction.

Gravity in the universe, Einstein indicated, causes space-time to be curved. Consequently, the space of an event in someone's life occupies no precise position in time within the perceptions of the unconscious, so the place and time of a current response can be mixed psychologically with earlier and future places and times in one's life.

Remember that in the contracting universe that will ultimately occur (Hawking, 1988), humans will know the future, but not the past. This contradictory potential underlies the human motive and our emotional interest in constantly wanting to connect with the future. For some, looking ahead can be frightening, especially if the imagined future bodes ill. Conversely, good futures that do not come true can be disappointing, and in some instances even arouse destrudo. Nevertheless, being pleased when wishes actually occur reinforces the futuristic element of construdo.

Unfulfilled anticipations from infancy and childhood can lead to discouragement of adult construdo development, as can parental failure to reinforce their child's construdo fantasies and plans. Parental disapproval, belittling, or attacking fantasies or plans leads to the worst outcome, causing excessive development of destrudo in some people, a pathological beginning for this drive – a direct byproduct of underdeveloped construdo or overdeveloped destrudo. Destrudo is also abetted by absence of sufficient preparedness along with a corresponding absence of sufficient practice of whatever needed to be practiced.

Gravity is the infant's most obvious physical connection with reality, from the time the infant is held upright, first tossed in the air, and knows it always comes back safely into its parent's arms. Gravity is a consistent kinesthetic experience for newborns, giving them their first glimpse of the planet's reality, the surroundings in which we live, a reality with which construdo can flawlessly and accurately combine the baby's mentality. Gravity's influence is felt on our language when we say "she has her feet on the ground," meaning that woman is well grounded in reality. Gravity affects the absolute unconscious (abs-unc) in causing it to sense, or to intuit, that other forces of physics act upon humans.

This abs-unc is the human unconscious psyche that develops within the fetus from three months to nine months in utero, and continues developing during the first six months after birth. During this development the abs-unc perceives the surrounding environment very much through unconscious visual comprehension – the "undifferentiated chaos" of Piaget (1954). TDT thinks differently. TDT is quite distinct from that part of the unconscious that develops out of the abs-unc as the drives differentiate into the drive-differentiated unconscious (DDU). DDU evolves from six to nine months in utero and in the first six months after birth. According to TDT, influences from the universe are registered in the abs-unc.

The forces of gravity provide humans with a grip on psychic reality as well as physical reality, which is also registered in the abs-unc (absolute unconscious) and then influences the DDU (Drive Differentiated unconscious). Moreover, the influences on the abs-unc from the operations of subatomic particles within the atoms of our brains – of confinement, or tightly pulling and holding particles together – cause the psychoanalytic fusion or "condensation" influence in the abs-unc. The smashing collisions of particles into each other and then the creating of new particles, and the changes in particle identity, cause the psychoanalytic displacement influence in the unconscious. All are registered in the abs-unc, which then influences the DDU. The timeless, spaceless condition of subatomic particle operations, exactly what Freud found to be the nature of the operations of the unconscious, is stored in TDT's abs-unc.

The DDU shows that drives operate more and more independently of each other, seeking differentiated goals. Construdo seeks to combine and connect people with their surroundings. Destrudo seeks to break apart these connections. In libido the organism seeks to reproduce itself. Once reproduced biologically, the forces of physics cause the resultant person or living thing to follow the construdo and destrudo drive aims and derivative aims.

The future lies in the space-time direction of construdo aims as demonstrated by human wishes, daydreams, fantasies, and future plans. It is in construdo's always planning ahead, in the wishes behind night dreams. This understanding includes responsiveness or orienting oneself to potential influences from forces of physics that may come into play billions of years from now, as the universe collapses into itself. As expressed earlier, at that time humans will know the future, not the past (Hawking, 1988).

One translation of the potential future influence on the DDU from the abs-unc is to anticipate or fantasize about what our future might be. Construdo creates these fantasies in its thrust forward in time, when good or bad destrudo things are anticipated. We can anticipate or plan for bad creations of construdo coupled with self destrudo. Individuals combine these fantasies with their self-destrudo concepts, creating both positive and negative anticipations for themselves.

Chester (1978) noted the change in identity of a subatomic particle. This is akin to the energy and energy directions of the drives supplied by the human unconscious. Hence, a drive or drive combination can reverse identity (e.g., from sadism to masochism), involving combinations of the three drives and a change or reversal of the object of the drive from an external object to the self. It might involve loving a person, then hating the person. Such reversals are confusing to an infant trying to gain a consistent perception of its surrounding reality with which it seeks to be combined. The infant's construdo wants consistent love from its parent. Changes or reversals in such love cause the infant to change positive self-construdo into negative self-destrudo.

A second translation of these influences from the DDU into the conscious psyche – developing construdo (the ego) – is an influence on the conscious psyche to return, in part, toward undermost energy origins of significant happenings in the past. This is a broader, deeper, explanatory concept of the repetition compulsion in our psyche, stemming from the fact that in billions of years the universe will return in space-time to its beginning stage or origins. In the Big Crunch, the universe will collapse totally into itself (Hawking, 1988), just as stars have in parts of the universe collapsed into themselves to form black holes. The mental energy of the repetition compulsion in the psyche to return partially to earlier ways of reacting and functioning mentally is parallel to the way energies of psyches operate in the universe. TDT postulates that what happens to parts of a system will ultimately happen to the entire system; it's a foretelling of where these forces of "returning to undermost energy origins" will take the whole universe. TDT deduces this conclusion from observing how mental or psychic forces operate.

When unconscious processes' underlying drive reversals invade human consciousness, they are often re-repressed, because they threaten to make a person's logical thinking psychotic, disordered, or neurotic by going backward in time to a self-directed stage of a drive in the abs-unc itself. The unconscious is incompatible in its operations with consciousness, because of its antithetical influences from different realities, namely, from the universe and subatomic particle influences on the human psyche.

Human conscious reality tells us time only moves forward. Hawking (1988) describes this effect as a fact of the reality around us, because people observe that things move on our planet from ordered states to disordered states, and there are many more disordered states than ordered states: it's the second law of thermodynamics, or the principle of entropy, which indicates an increase in the passage of time to disordered states. The passage forward in time, like the direction of the expanding universe, is registered in the unconscious, telling us time is moving forward. As discussed earlier, in the very distant future, humans will demonstrate the capacity – in a collapsing universe – to relate to the fact that time is moving backward. Then, according to Hawking (1988), they'll be able to tell the future. At the present time this ability is already a potential from the unconscious. At times we use it; most times we don't.

The nature of subatomic particle operations has resulted in quantum mechanics for predicting the likelihood or probabilities of particle movements and events, as well as in understanding the uncertainty principle. As a result, construdo is influenced psychologically by feelings that events in life are often unpredictable and uncertain. This is one root of psychological anxiety and insecurities about outcomes in our lives. It's a result of construdo's apprehension about joining surrounding events or the imagined worst outcomes. Such imaginations can fill the infant with dread. This is the experience of anxiety. This is the "exclusion principle" of physics, for the atom indicates that two matter particles of spin $1/2$ cannot occupy the same position and have the same velocity within the limits of the "uncertainty principle."

In terms of the drives, the exclusion principle indicates that the drives actually fire sequentially, not concomitantly, as is traditional in the psychoanalytic concept of fusion. Yet the drives may be combined by construdo so that they discharge both in close proximity and, alternatively, alternating in a scintilla of a moment.

Individuals cannot focus on arousing one drive aim toward a given person, in the belief that it will also cause a second drive aim to also discharge toward that person. Individuals discharging object-construdo, constructing a relationship and feelings for another, should not assume that this will also result in heightened genital libido attractions between them. The arousal of one drive does not automatically cause the arousal of a second drive, because the drives do not fuse. Rather, they alternate in their

185

discharges, which could turn them away from each other; they get disillusioned.

Implications of this exclusion principle also mean that two drive stage aims (either from the same drive, or from two different drives) cannot be discharged toward the same object in a given place, at the same speed and time. One arouses a person before the other.

The first arousal has primacy over the second, and these sequential arousal times cause variation in the responsive behavior of people, but always within the limits of the uncertainty principle. Consequently, the resultant responsive behavior can never be precisely predicted. But psychologists, in wanting to be scientific, have thought they should be able to make such predictions. TDT has established that they cannot. That is because of the many ways the uncertainty principle comes into operation and influences the outcome of the divergent forces. A second implication in the drives' unconscious is that a drive will not discharge unless it has an exit and a return pathway for replenishing itself in its unconscious origins. This is the "replenishment principle" of triadic drive theory.

FREUD'S CONCEPT OF A CONSTANT DRIVE INFLUENCE

Freud (1915/1957a) asserted that an instinct or drive resulted from a constant influence or pressure on the unconscious psyche. Individuals are unaware of this influence. Therefore, for example, the influence of gravity on the abs-unc and the DDU creates an influence on the conscious psyche, the ego conscious construdo. Gravity, one of the forces of physics that creates the construdo drive, fulfills Freud's notion of a constant pressure, because gravity influences everything, everywhere on our planet, at all times: it is a constant force.

No one is ever aware of the constant pressure of the gravitational force of physics on their unconscious psyche. Nevertheless, gravity binds all of us to the absolute space-time concepts of our planet. However, these space-time concepts can be disturbed by the influences from within the atom that show timelessness and spacelessness when differentiated in the individual's abs-unc.

INDICATIONS IN DREAMS OF TIME DELAYS IN
UNCONSCIOUS OPERATIONS

The concept that the three drives of the Triadic Drive Theory have separate and sequential influences can be better understood by considering differences in times when the drives entering the dreams. While the influences of all three drives appear in dreams to varying degrees, their aims are all subject to distortions. Distortion of the dream thoughts leads to the construdo creation of a dream.

An infinitesimal time must elapse before a second drive influence can enter a dream. Time is a factor creating distortion in dream content; that is, time must elapse before the second drive's entrance into the dream, because of the exclusion principle in subatomic particles' operations. Two drive goals cannot be at the same place at the same time. This result follows a sequential order of time and is one factor causing transformations and distortions in the latent dream content. Space or place is the second factor. The content formed by the abs-unc (absolute unconscious) influences the DDU (drive differentiated unconscious) in its creation of the manifest dream content.

These conclusions are different from the fusion idea of dream images. Fusion results from the condensation process, in which two or more dream thoughts and images become condensed into one dream image. But insofar as the perception that the inputs from two drives create one drive image, – the input is sequential, not fused. Thereafter, fusion often occurs, as in instances of symbolization, where one dream image represents a host of dream thoughts and ideas. According to TDT, all these fusion phenomena stem from the confinement property in subatomic particle operations and its influence on forming the abs-unc and subsequent DDU dream images. The confinement property explains why latent dream content can be revealed and unraveled by associations to the distortion of the drives' aims in the dream. Remember, the unconscious space of an event and its time are not represented in a dream.

OTHER REALITIES CAUSING DRIVE PSYCHOPATHOLOGY

Drive regression is the result of the "principle of returning toward undermost energy origins." A current stimulus may be reached as if it were an older, similar stimulus, first presented to the individual in times past. The individual responds in past ways. The result is an emotional reaction that has lingered in the individual's mind from times past. It is reactivated by the current similar stimulus configuration. These older emotional reactions are then called forth, or activated, by the present similar stimulus configurations.

Such emotional reactions are displaced in time to the present situation by the abs-unc. If the regression goes farther back into the abs-unc in the psyche, then the individual totally responds to the present situation as if it were the earlier situation. A wide array of responses can occur, because earlier responses are not connected with how the individual presently responds.

The drives emerge into the DDU in the following order: first construdo, next destrudo, and finally libido, occurring in utero and during the first six months of life. Regression can go back to the first grouping of the unconscious's three undifferentiated drives, a psychotic condition: to pure abs-unc, or to the advanced DDU, as well as alternations between the two – still a psychotic condition.

Individuals can experience mixed feelings and confusion about which drive should predominate in reaction to the presenting stimuli. Often they may respond with contradictory drive reactions. When they express these confusions they may be judged to be psychotic. Their present time orientation is displaced to former, earlier times, and they react as if the earlier situation is presently occurring.

Personality disorders follow to a lesser degree from later organi-zation of the distinct drives from six to eighteen months of life, when the influences from other realities are lessening from the abs-unc. After eighteen months of age, when the drives are clearly distinct in the unconscious, regression to this level of the DDU of organization most often results in neurotic conflicts in the drives.

DRIVE REVERSALS AND OTHER REALITIES

Freud considered the change from sadism to masochism a drive reversal. He thought sadism, the aim to torture, was being turned onto the self in masochism, or a change from object to self. It was also a change in energy output from active to passive (Freud, 1915/1957a). By this time, Freud (1905/1953b) had already presented his concept of the libido drive from the unconscious. In 1915, he considered aggression to be from the ego, or what he called the "self-preservative" instincts. The aggressive component of sadism and masochism in TDT is from destrudo, rather than any emotional component of a self-preservative ego instinct, to protect and defend oneself. Freud (1920/1955a; 1930/1961a) did not put aggression on the same level of the unconscious as libido until much later. TDT does so from the start.

TDT considers the change from sadism to masochism in the same individual as a change from object-destrudo to self-destrudo, and from object-libido to self-libido. Freud (1914/1957c) called the latter *autoerotic* – namely, the self is the object of one's own libido. In TDT both drives are connected by construdo. Construdo connects the object aims of sadism, then disconnects them, and finally reconnects them as self-aims to produce masochism. Remember, TDT asserts that the drives discharge sequentially, not in fused combinations from the unconscious. Hence, in TDT self-destrudo and self-libido aims are combined by construdo from two distinct drive aims.

Masochism involves inflicting destrudo, then pain on oneself, and next deriving libidinous pleasure. TDT views this situation as involving two drives reversing themselves rather than one drive, as was Freud's (1915/1957a) view. Yet both drive discharges afford pleasure to the individual in the situation.

In later life, for example, males derive libidinous oral pleasure from nibbling on the nipples of a female lover or in biting on them – a pincer-object-destrudo action from which the male derives object-destrudo satisfaction. It is, nevertheless, what Freud called sadistic behavior. In Freud's conceptualization, when a male willingly submits to a female biting his cheek or lip – that is masochistic. In TDT it is self-pincer-destrudo combined with self-libido behavior. Sadism and masochism can be seen to go forward and backward or reverse themselves, as do subatomic particles and antiparticles within the atoms of our brains. Moreover, the actions are returning in each drive toward their undermost, self, uterine energy origins.

Each of the three drives begins in self-construdo stages: self-construdo and self-destrudo (combined with the self by construdo), then self-libido (also combined with the self by construdo). The uterine energy of self-construdo is the beginning of all sense of the self in humans.

From the drives in us, to the universe at large, we can discern the principle of returning to undermost energy origins. The forces of the expanding universe from the Big Bang's energy versus the gravity energy attracting bodies to each other will ultimately cause the contraction of the universe back into itself. It is a constant potential energy pull to return back to the original, intensely packed energy particle which theoretically will lead back to the Big Crunch (Hawking, 1988), which TDT speculates could ultimately be another Big Bang explosion as occurred at the beginning of the universe. That is what we mean by a return to "undermost energy origins."

Reconsideration of Freud's data shows that TDT's three drives would be involved in what Freud (1915/1957a) saw as a reversal of a single drive. The difference in viewpoint leads to very different conclusions. The three drives presented, plus the specific stages and aims for each, leads to seeing the data very differently than Freud did in 1915.

TDT perceives that the two psychic drives from physics often alternate in their expression, from construdo to destrudo, and back again. This is congruent with alternating forces of physics from all levels of the universe. That's the first alternation from the extremely dense energy particle that exploded into the matter of the universe in the Big Bang. It is in the galaxies pulling together, then apart, in multi-billion-year cycles. It is in our solar system's sun spots appearing and disappearing in eleven-year (or longer) cycles. It's in the earth's magnetic north and south poles reversing themselves in million-year cycles. It's in the subatomic particles as electrons changing into positions, then back to electrons, in a scintilla of a second. From our biology it's in the building up (anabolism) and breaking down (catabolism) processes in the cells of our bodies. All these influences give the unconscious its directions to alternate our drives.

Construdo disconnects the self, or part self, from destrudo aims and self-libido or part self-libido aims within individuals; then it reconnects them to object or part-object destrudo and libido aims. Thus, all three drives are involved in reversals of what was seen as a single drive by Freud (1915/1957a), in the reversal from sadism to masochism. TDT sees the reversal as caused by changes in the three drives.

The change from self to object or from object to self is underlain by the determining influence of subatomic particles changing from particle to antiparticle, or a backward and forward through time change of particle identity, as well as the change from the universal forces of expansion to the ultimate future contracting of the universe, the fundamental alternation of the universe, registered in the unconscious, and affecting the drives, causing them unconsciously in the DDU to reverse themselves, or alternate.

Reversal of the Drive Content

Freud (1915/1957a) observed a reversal from love to hate. TDT sees this reversal as involving changes in each of the three drives, rather than only one. An individual's construdo disconnects and reverses itself from its creative creation of an idealized loved person to constructing plans of destrudo vengeance for rejection – real or imagined – by this loved person.

Individuals falling in love often report feelings of being "knocked off their feet" in experiencing the loved person. Here, self-destrudo gives them this self-toppler aim and result. Individuals then either seek to object-destrudo topple the formerly loved persons from their position when love existed, or they want to hurt, through object-striker destrudo, these formerly loved persons and cause them the pain they now experience as their love has ended. Destrudo vengeance can last as long as it does because the loved person has been so extensively, lovingly, and positively fantasized over long periods of time in the created, idealized object-construdo connecting aims of this drive.

At the same time, libido may change its genital aim to an earlier stage aim of an anal retentiveness, causing the individual's unconscious libido to seek holding on to the lost, loved person. So, too, does the object-

190

destrudo's unconscious pincer aims. Both will give rise to various efforts to consciously control the person who is slipping away. Individuals' destrudo here will also seek vengeance for their wounded ego on being left.

Love lost also produces depression and feelings of mourning for the lost love. It's a different reaction to the loss than feeling hate and a need for vengeance. Individuals in this depressed state have planned their future on their relationship with the loved persons – a construdo fantasy. In defeat of their dreams they experience self-toppler destrudo. It is expressed in reactions leading individuals to feel that the ground has been pulled out from under them. They have been toppled! They no longer feel anchored in their environment. They no longer feel grounded (by gravity). Construdo has reconnected them with their self-toppler aim to beat themselves down to where they feel no connecting with love. Individuals in this mental state have lost for a time the construdo drive's aim to connect themselves with their environment. They contract into themselves, as gravity will recontract the universe in ultimately distant future times.

Under these conditions, individuals experience depression. They may even experience suicidal feelings (intense self-destrudo) because psychologically, they do not feel any belonging or connection with their world. They have no construdo impetus to plan for themselves in the future. They have no dreams or fantasies of a future. They feel pain and sorrow in realizing that their love fantasy did not continue.

They also feel guilt about not fulfilling their love fantasies. Guilt produces destrudo. They turn the aroused destrudo from this onto themselves, causing more depression. In these drive aims and direction there's a depletion of any positive construdo energies to connect themselves with the world.

Libido in this state is disconnected from all object-libido aims. It returns to the self-contactual aim, from uterine times, seeking contact with the self in sorrow and depression, as a way of yet feeling pleasure of a self-contactual libidinous nature, by contact or connection with oneself. It's pleasurable because it goes back to a place where the sense of self once resided. If nothing else, it provides leftover pleasure in being connected with oneself, alone as this might be.

This is an instance of the universal forces causing the individual to return to the self's undermost energy origins. It is also a shift to a backward-in-time direction, or an alternation that the drive can achieve because of the influence of subatomic particles. The individual changes in identity, from being happy to being depressed.

Artistic and creative individuals, who are functioning from their object-creative aims, sometimes report or exhibit opposite feelings of self-destrudo aims; that's toppler, striker, and smasher self-destrudo aims. Their elation over their creative productions are accompanied by wanting to participate in their own self-abuse – heavy smoking, drinking, or drug abuse, or some other way of knocking themselves down – self-toppler and self-striker destrudo.

Of course from a TDT perspective, self-destrudo aims are an outgrowth of the influence from within subatomic particle operations on the unconscious from particle annihilation. When subatomic particles collide and annihilate each other, new subatomic particles are then created. Psychologically, then, the creative aim is linked unconsciously to the destructive aim. It matters not which process comes first, because in unconscious influences time is not a factor and is not logically sequential.

The libido, one pleasure component of the creative process in people, follows from the contactual stage of libido, first seen in the third trimester of fetal development. Individuals here feel a connection with themselves or with the "center" of themselves. It's a deeply felt construdo self-connection, causing great pleasure in the self. It's the pleasure from the self-sensed ability of creation. They may feel, on the other hand, that they have lost such connections; they have lost their connection with themselves. A transitory lack of motivation, despair, and depression follows. Once they regain their center, happiness can reappear.

Hence, in the three reversals described in this section, failing construdo connections with external objects or persons or with oneself result in lacking feelings about fully continuing one's life, and participation of oneself in the surrounding environment can cause depression. It's the result of the construdo drive's temporarily seeking to disconnect with the surrounding world and not seeking to reconnect.

Healthy, nonpathological existence depends more on the healthy operation of construdo than on the other two drives. When self-construdo is involved in a bond with a loved person, a lifelong friend, or with a given locality (e.g., a favorite street in New York City, or an apartment), a cherished event (e.g., a special interaction with a parent or a scene in a movie), breaking the connection can be truly traumatic. It's tantamount to giving up a valued part of self; it happens when a loved person dies or moves far away. In these instances, construdo does not yield easily to the disconnection. Depression may be the reaction to these losses, which may persist for years.

Harmony with the Universe

As indicated, construdo is the primary drive – in its continuation – for a person's mental health. In humans this drive is derived from the forces of physics that bind and hold all matter in the universe together, from galaxy clusters down to the atoms within us. This construdo drive holds together our psyches, conscious and unconscious, just as the strong and weak nuclear forces hold our atoms together, and the electromagnetic force holds our molecules; just as substances on our planet together, just as gravity holds us to our planet and our solar system and galaxies together in the universe. It is the drive in our psyches that allows us to be in harmony or congruent with the rest of the surrounding vastness of the universe.

Describing Our Drives
by the Other Realities of Space-Time

Adding construdo to the dual-drive theory, creating a three-drive theory, not only requires that the past of the drives and their vicissitudes, development and fantasized unfulfilled aims, fixations, repressions, and sublimations should be taken into account to understand the psychological present, but that the anticipated and fantasized future, the certainty of construdo and in general destrudo and libido, should be taken into account to understand where the drives are attempting to lead us.

Construdo gives us a more complete psychological picture with the introduction of the anticipated, fantasized future as a determining factor of behavior. It also necessitates a science that can deal with predicting the probability of future events. That is just what quantum mechanics does.

If we can determine where a drive's future fantasy aim began, we still may not be able to determine how fast it will move into consciousness (or when). If we observe such a fantasized aim of a drive moving into consciousness from the unconscious, we still may not be able to accurately determine exactly where it began. This is the "uncertainty principle" in operation.

Modern physics gave us the four-dimensional reality of space-time and the realities of the universe. What is true in the universe and its forces will be true of infinitesimal parts of the universe, such as with respect to human beings, and how forces within their brains operate – namely in the atoms of their brains, causing their drives. We consider the data of psychology in terms of our conscious reality, a three-dimensional reality. That is the visual, tactile reality of our planet. Gravity is the force to which our perceptual senses are most responsive.

Galileo, Newton, and Hawking asserted as founders of 20th century modern science, distinguished from 19th century science, philosophy and religion (Hawking, 1988). It is their works that have shown the nature of this force of gravity in the reality of our planet, which is registered in our unconscious as the force that made us feel "grounded" in our uterine existence and the first six months following birth. Gravity gives us our kinesthetic sense of being connected to our planet. When we say, "a person has his/her feet on the ground," psychologically we mean this person is in a positive contact with the consciously perceived reality of the planet. Yet our other realities are from the universe, a four-dimensional reality, and the opera-tions of subatomic particles in our brains have an influence with respect to TDT, on our unconscious.

It should be noted here that present-day string theory describes five to seven additional dimensions, and p-brane theory indicates up to ten or eleven dimensions (Hawking, 2001). Both theories suggest added folds in the subatomic particles as causing the increase in dimensions. At this time, however, I do not think our conscious minds respond to these additional dimensions. When more details are forthcoming from physics about these dimensions, it can better be determined if they have any effects on our unconscious.

Earth's reality connects us with all that we understand around us, the world we perceive with our senses. This is the first reality to which we are responsive consciously, giving humans their first sense of being "in the world." Whatever that means, humans search for it throughout their lives. But in TDT thinking, the other influences from the universe and the operations of subatomic particles in their brains are also extremely influential in their mentality.

CHAPTER 16

THE CONCEPT OF CHARACTER TRAITS

FREUD'S CONCEPT OF CHARACTER TRAIT FORMATION AND TDT'S ADDITIONS

In his paper called "Character and Anal Eroticism," Freud (1908/ 1959a) advanced the concept that components of the drive-aims of libido could reemerge from the unconscious organized around libidinal components as character traits. Here he cited traits of being "orderly, parsimonious and obstinate" (Freud, 1908/1959a, p. 169). These traits are the result of libidinal anal retentiveness from the child's toilet-training times – that is, between two and three years of age. The struggle between the parent and child to do what the parent wants results in orderliness, oppositionalism, and problems with authority figures. In "On Transformations of Instinct as Exemplified in Anal Eroticism," he further advanced the concept that transformations of components of libido-aims in the unconscious could result in such character traits (Freud, 1917/1955b).

After completion of the developmental stages of libido (which Freud put at five years of age) and the developmental stages of destrudo and construdo (which TDT puts at six years of age), Freud found that shame and self-disgust plus learned morality become influential in inhibiting and repressing a drive's component aims. That is, the superego or (in TDT) the extension of construdo is focused and connected with self-evaluation and self-esteem. When inhibiting, repressing influences are bypassed in the psyche, the component-aims continue into consciousness as character traits through continuation or prolongation of the original component-aim energy. The process is directed by construdo energies.

The process may also involve breaking away or separating the component-aim from environmental attachments. This is accomplished by destrudo influences. Hence, an aim is held together and kept unto itself by the subatomic property of confinement. It is tantamount to an unconscious condensation and timelessness regarding the aim, keeping it moving on and on, without regard to the passage of time or place.

The component-aim may be displaced in the unconscious to another place in the unconscious. This may cause slightly different aims to emerge, resulting in slightly different character traits. The drive component-aim may be displaced in the drive to another drive.

Triadic Drive Theory (TDT) will show that destrudo and construdo's various stage-aims are also involved in trait formation as found with libido. Parsimoniousness, in addition to the anal holding-onto-feces aim, is also an object-pincer destrudo-aim resulting in the controlling of one's possessions, as well as construdo plans or aims to construct and create what the situation should be. Obstinacy involves a toppler object-destrudo competition aim of who is going to dominate the toilet-training situation, the parent or the child. Obstinacy is an aspect of construdo to be connected in the situation to future intentions. It is also achieved by pincer object-destrudo efforts in the battle between parent and child as to when and where the child will defecate.

We observe here libido occupying a middle position between destrudo and construdo aims. Libido in some situations will determine how these drives will unfold. However, destrudo and construdo are the drives that cause the determining influences in most of these decision making situations.

Displacement of a drive's component-aims may also result in a sublimation of the aim, or it can be reversed by the DDU (Drive Differentiated Unconscious) into a socially approvable new conscious aim. These sublimations and reversal transformations turn shame, disgust, and moral self-condemnation into aims that will gain social approval and approbation. Such reversals are the work of the abs-unc (absolute-unconscious) processes on the DDU. Reversals, in TDT terms, stem from the alternating influence occurring throughout the universe.

What then emerge in an individual's psyche – unconscious and conscious – may well be creative productions or expressions as sublimated character traits that the individual has an affinity for. Here the subatomic brain influence further involves an identity change with the alternation. These identity changes may involve furthering in themselves some special talent or ability that they combine and connect within themselves, and for which they do not feel fearful, insecure, anxious, or uncertain in themselves.

The talents may be in physical, visual, or auditory expressions of the self. There can be combining of gustatory and olfactory senses, as in cooking or being a gourmet. The consistency or unevenness of a food when chewed, or the sense it evokes in an individual, may connect it in one's mind as a certain food pleasure. Tactile senses may be involved. In any of these sublimated talents an original drive-aim or combinations of aims can be moved or displaced in the DDU to a new place in the psyche. Thus, new character traits can emerge. It is another basis for identity change.

When a reversal of a drive-aim component is involved because of superego self-condemning factors, causing a reversal of the aim by DDU forces, but the individual senses no special talents or means of expression for solving the drive dilemma through sublimation, then a reaction formation against the original drive-aim component is likely to occur. The individual displays expression of the opposite of the drive-aim component. The universal alternating influence causes this drive alternation or reversal.

A displacement of an aim, following its reversal, may also occur. In addition, the self-condemning superego influence in the DDU arouses self-toppler destrudo, which will move the ideation of the original drive aim component backward in time, before it existed in the DDU. Self-toppler destrudo moves the original drive-aim component back into the abs-unc, where processes may reverse it. It is this way in psychosis.

When the reversed aim comes back into the DDU, following the alternating processes between these two levels of the unconscious, construdo combines with the newly emerged aim, carrying it forward with its psychic energies toward the new, fantasized, positive aim in times ahead. The interaction in the DDU (Differentiated Drive Unconscious) between destrudo and construdo, and in subsequent space-time, parallels the changes in time directions that originates in the abs-unc (absolute unconscious). It is seen in psychosis.

RETREAT, REPRESSION, AND REGRESSION

Although a drive may solely contribute to trait formation, the more usual situation is that the drives conflict with each other. The resultant defensive operations against one drive that is unduly predominating are seen in retreat, repression, and regression. Lapses in the development of the drives occur when superego influences, drive conflict, drive trauma, or drive arrest, stop a drive's forward development. We look now at these drive conflict resolutions.

Retreat

The first indication that individuals have a drive or drive-derivative conflict can be observed when they retreat from a confrontation between destrudo strivings and construdo strivings, and derivative conflict strivings of the two drives. Such individuals draw back from their aroused destrudo, and it moves them backward in time, before they were led into the confrontation. They do not want to connect with the confrontation of their own or the other's destrudo aims. They resist in themselves; and with the drive of the other's – they are in conflict and discomfort.

These individuals retreat to avoid a conflict they feel will put them on untenable ground and cause them anxiety. Their destrudo or construdo drives can remain free and unencumbered and continue in natural development. However, if retreat from drive conflict occurs repeatedly, becoming a pattern of behavior, then the unexpressed drive energy may simmer in the DDU. Such individuals retreat from the possible wish for the unexpressed drive energy to boil over and show itself. The wish (construdo aim) can build up after being held in for years.

Such a conflict of the unexpressed drive energy to show or not show itself might also arise between object construdo, libido, and destrudo, when feelings of sexual union, excitement, passion, are threatened by angry, hostile, hurt feelings toward the object. A conflict may arise between destrudo, libido, and construdo when so-called sadistic object strivings are the way one wants to be connected with the object. Sadists connect with the opposite sex in ways that hurt, torture, humiliate, and disparage. They gain libido satisfaction and pleasure from such treatment of the other person.

Repression

When the drives or drive derivatives conflict, and parts of the conflict are intolerable to one's conscious mind, it will be blotted out of consciousness. Pushing the drives' conflict back in time are forces of destrudo. The unexpected construdo conflict arouses destrudo. This separates construdo's striving to stay connected with the two conflicting drives from destrudo's striving to break them apart. When destrudo prevails, it pushes part of the drive thoughts out of consciousness.

Part of such drive thoughts are pushed out of consciousness by smasher thought destrudo aims. Such drive thoughts are repressed – blotted out of all awareness and memory in consciousness. These drive thoughts are completely forgotten. Yet the drive's aims and strivings can continue to influence the conscious mind. Psychic remnants can remain from the two conflicting drives. When libido is involved between construdo and destrudo conflicts, libido then can become the central drive around which the other two drives resolve their desperate influences.

However, libido may not be the all-binding drive and unconsciously all-uniting drive between couples that Freud (1905/1953b; 1923/1961c) believed. For example, partners in a couple will experience the arousal of destrudo when their construdo expectations (or fantasies of what should occur between them) do not occur.

Construdo's normal forward direction in space-time is reversed by the backward energies of destrudo from the DDU (Differentiated Drive Unconscious). Once set in this direction by the pull of the reversing energies from the abs-unc (absolute unconscious), complete repression of drive aims results.

Regression

When fragmented and clustered drive energies and aims enter consciousness from fragmented drives in the DDU, and drive conflict is repeatedly encountered in conscious reality without being dealt with, then regression of the fragmented drive energies occur. The persons involved sense that their pertinent drive energies are inadequate for the situation. They simply cannot deal with the situation confronting them. Their destrudo smashes and strikes out the conflicting ideation, pushing it backward into the DDU and further into the abs-unc. It is hypothesized that how completely these mental operations carry the ideation determines the depth of the psychotic regression defense.

In reality, these people contract and withdraw back into themselves. In Triadic Drive Theory this is an influence from the abs-unc on fragmented DDU, derived in our TDT from what is considered to be the influence in the universe that ultimately will recontract the universe back into itself (the Big Crunch). Because of this universal contraction influence individuals unconsciously move backward in space-time, where they confront reality as if it were events (space) from their past (time), and which have little or no relationship to present events. They go backward in space-time as if this were the direction of normal development. This backward development, or progression of destrudo, is called regression.

When their psychic regressed drive energies begin to alternate between the DDU and the abs-unc – it results in psychosis. The greater the frequency and intensity of these alternations – the more severe is the psychosis.

THE SPACE-TIME FACTOR IN DRIVE CONFLICTS

Drive conflicts can cause individuals to escape from the present conflict by reacting as they did to a similar event from an earlier space-time in their lives, or to a future space-time, as in a fantasized event, away from the impact and intensity of the present happening. These defensive moves in the psyche are to avoid hurt and pain aroused by the present event. Destrudo takes individuals backward in space-time, paradoxically, to similar painful events that were dealt with in some surviving way. Individuals tend to repeat how they dealt with a conflict in the past, as they cannot see it realistically in the present. They are glued to their past reactions.

On the other hand, construdo may take individuals forward in time to future fantasies that pass over the present conflicting event, taking these individuals to a better space-time, where they deal with the conflict successfully in fantasy. It is their hope, but it seldom happens. They tend to repeat their responses from earlier times, while yet believing they'll do something different. A person's time sense is related to their drives' conflicts and resultant new drive directions in their psyche, in addition to the time on a clock or calendar.

A drive conflict in the present can also be avoided by construdo fantasies of how individuals wished the event might have happily unfolded. They mentally picture a happy ending. Such fantasies are commonly observed when individuals rethink events in which they are dissatisfied with their participation. They think, "I should have said... I should have done this or that – if I was quicker..."

The fantasy in young students of going to college will in coming years determine their behavior in college. The desire to marry a certain person can determine one's behavior toward that person for years to come, both consciously and unconsciously. The influence of the past on the unconscious is the focus in the section to which we now turn.

CONFUSION IN COUPLES ABOUT THEIR DRIVES

The effects of the three drives in couples' relationships and their sex life can be observed continually in clinical practice. How often does a therapist hear, "We had a great sex life, until we got married," or "until we had a child," or "until I had an abortion," or "until we began fighting about everything," or "until I got so mad I could never forgive him (or her)."

The first three statements refer to construdo fantasies or its plans about how in the future the couple believed they would be combined and connected with each other, as they were before these events occurred. It might be the type of relationship they had constructed in their own minds, without discussing it with each other. They both had similar fantasies, although they probably never stated these fantasies to each other. But when their separate fantasies about their relationship failed to materialize, their destrudo was aroused toward the other person, and they express hostility toward each other without ever knowing its true cause. What they think is their real irritation – some minor differences – is rarely the real cause.

The final two statements refer to aroused mutual destrudo aims that were never satisfied and could only push the couple more apart. Libido between the couple often does not and cannot bridge the gap with these construdo failings and the aroused destrudo results. We have here two very different, conflicting drive sources: libido versus construdo and destrudo, the first derived from biology and the last two from physics.

One consequence is feelings of alienation within oneself because of these very different drive origins. Humans are left with puzzling feelings coming from different parts of themselves. The resultant alienation caused by these two very different sources of drive energy divides them within themselves. One part of them feels alienated from the other part of them. They can sense both parts in themselves, and they are puzzled about such strange feelings in themselves, seldom experienced except in drive conflicts. Drive conflicts create feelings of estrangement and alienation in people.

Libido will succeed and flourish if the couple heeds the libido aims, satisfactions and fantasies. If they expect that libido aims are similar to construdo aims, difficulty with one of the drives will surely occur. Little wonder then that the actual reported experiences of couples do not turn out as common wisdom expects. If only one drive (libido) was involved in the passion of the relationship, while only construdo was involved in the creation of the love relationship, then the cause of their difficulty in becoming a couple can be understood as a construdo failure. They do not understand, however, that this second drive (construdo) is predominantly involved in what they are trying to put together with each other – that is, in working out and planning details of the life they want to spend together. They usually expect that the passion and urgency of libido between them will impel them to want to be together. But, actually, it doesn't.

In truth is – in the reality of the DDU (Differentiated Drive Unconscious), a different situation occurs. What's seen on the unconscious drive surface belies the reality of the unconscious drive situation. Libido joins with either construdo or with destrudo in the basic, fundamental unconscious drive conflict between construdo and destrudo. The constructed relationship will then either tilt toward passionate coupling or loving togetherness, or toward a tendency to constantly breaking the bond apart. The latter ends the chances of a long-lasting, continuous, and happy relationship.

CHAPTER 17

FORMATION OF DESTRUDO CHARACTER TRAITS

DESTRUDO-DERIVED TRAITS

It should be understood that specific destrudo stages and aims were not presented by Freud (1920/1955a), although he did initiate the concept of the destructive drive in *Beyond the Pleasure Principle*. He believed it was a biological drive; Triadic Drive Theory (TDT) does not.

Let's examine characteristic traits resulting from the component drive aims derived from the forces of physics. We will consider them along the three lines Freud proposed for character-trait formation from component libido drive aims: prolongation or continuation of a drive aim, sublimation of an aim, and a reaction formation against a drive aim (Freud, 1908/1959a; 1917/1955b). For the sake of brevity, not all possibilities will be presented in each instance.

The Break-Apart Self-Destrudo Trait

When the break-apart self-destrudo aim from early uterine times continues into childhood, adolescence, and adulthood, an individual may show the character trait of construdo, connecting with and sometimes constructing situations that tend to pull or tear the individual's lives apart. The extent to which as embryos and fetusessuch individuals resisted and struggled against a spontaneous abortion during the first six months in utero will influence the intensity of this trait, as will the extent to which the pregnant mother was conflicted about her pregnancy; that is the construdo bond she shares with her fetus: wanting the child, having mixed feelings, or not wanting the child at all. The mother might have been afraid of pregnancy and/or giving birth. TDT considers the possibility that the fetus senses these deep feelings, fears, and conflicts of the pregnant mother and ultimately can feel the same way toward itself later in life as an adult. Thus, such an experience may stimulate severe periodic self-destrudo feelings.

Sublimation of the Break-Apart Self-Destrudo Trait

This trait may be seen as a sublimation, in the case of a sculptor who conceived of an art object being broken apart then put back together again in his sculpture. The breaking apart of an object was unacceptable to his own superego's standards and society's standards. Then he created an object in his sculpture that would be acceptable to society. He put the object back together artistically, in his creation. His destrudo broke down a reality and then construdo reconnected it with a new reality that he created. This is a sublimation of his break-apart self-destrudo aim. Here, the trait is an alternation of the aim.

A Reaction Formation Trait against Break-Apart Self-Destrudo

Without the above artistic talent, some individuals might find their abs-unc (absolute unconscious) alternates their drives' aims in their DDU (Differentiated Drive Unconscious). Reversed, the impetus to break themselves apart unconsciously changes to a conscious impetus to put others' lives back together. It is seen in the drug-rehabilitation counselor who helps reclaim people from efforts to destroy their own lives. This is a reaction formation or a displacement from self-destrudo to self-construdo through the counselor's efforts. It's a displacement by the DDU in moving the influence from one drive to another. The uncertainty principle governs the outcome of these influential factors in terms of what will be observed as the ultimate outcome.

Defective Part Causing a Self-Destrudo Trait

The sensings from uterine times of a defective part causing self-destrudo, which might have led to the death of the fetus after birth, can continue as a trait in later life. This self-destrudo is pulled together by construdo energies that construct a trait of being overly self-critical, too perfectionistic and unrealistic about oneself. When these individuals sense a fault or shortcoming in their personality, they see themselves as completely worthless and "no good," based on a single factor from very early in their lives. Thus, we propose they are contracting back into an early uterine sensing of themselves. Every time they encounter a fault within themselves, they denounce themselves in a depressive fashion.

We also propose that the extent, to which these people overcame the possibility of an organ failing them in utero, in the third trimester, will influence the intensity of the formation of this character trait. So, too, does the extent of the pregnant mother's construdo-linking connection with the developing baby within her, or her desire or contempt for her developing fetus. The stronger these two factors (possible organ failure and the construdo-linking connection) are in a negative direction, the stronger will be this excessive, depressive self-fault-finding trait. This amounts to excessive self-toppler destrudo behavior. Such behavior is reinforced by disappointments people have with themselves toward any hope or idea their construdo sets out for them. They lack positive self-esteem. Unconscious self-condemnation, in the DDU, is what they are often left with. It can result in depressive states.

Sublimation of the Self-Fault-Finding Trait

Sublimation of self-fault-finding is observed when people are determined to develop and expand to the fullest any talent they possess. Construdo enters the picture here in terms of how individuals begin to fantasize or plan connecting with others in the world by creating things that might captivate others. If they are successful in this construdo aim, it increases their self-esteem. In such cases individuals seek the admiration of others. They work hard to get it by changing themselves in ways to which others respond positively. Their construdo self-connections are becoming congruent with this goal. Working the construdo aim is a repeated under-taking that captures their motivations, in large measure, until they achieve success with others.

A Reaction Formation against the Self-Fault-Finding Trait

A reaction formation against the self-fault-finding trait is seen when individuals strive to intensely develop whatever talents they find in themselves. Note that not all efforts to develop one's special abilities are a reaction formation against sensing self-defective aspects. The difference is in the intensity with which these are pursued toward perfection of possessed abilities despite all obstacles. Such persons refuse to be "done in." They will not be defeated. They will not be bested. Rather, they'll best all others. The process may reverse poor self-esteem; coupled with positive self-connected, constructed and created construdo, they can become so intensely focused on a possessed ability that they become world-class tennis players, ballet dancers, violinists, physicists, surgeons, writers, chefs, generals, or dress designers.

The continuations and prolongations of uterine aims following birth become more complicated as life goes on, but the mental mechanisms or processes of these transformations regarding self-fault-finding traits remains the same. These are in the workings of the unconscious in accord with Freud (1900/1953a). TDT adds the information and proposes that these influences or processes were determined by the operations of subatomic particles within the atoms of our brains. In other words, process of expansion derived from the Big Bang and the concomitant forces of gravity that would pull the masses of the universe back toward each other possibly creates an influence on the drives, causing them to contract back into themselves in their operations. So, for example, libido can cause strong object attraction, while construdo expansion influence can be followed by a cessation of any such interest or a disconnection by construdo. Thus contraction back into oneself by the drives may follow.

Destrudo can change with advancing age from expansively fighting about everything to "mellowing out" and pulling aggressive aims back into oneself. The time frame here may be over an extended period. Similarly with respect to time frame, construdo can vary at different times in our lives. We can expand by reaching out to make connections with others, or we can disconnect ourselves from the outward expansion of our drives.

There are many possible character traits that can form from the prolongation or continuation of the energy from all of the various drive aims, their derivatives, sublimations, and reaction formations against these character traits. This chapter outlines the major traits that result from these factors.

After birth (during the first six months of life), the DDU (Differentiated Drive Unconscious) continues to become differentiated in the new circumstances of life outside the uterus. Objects or people, the environment, its events, stimulations (auditory, visual, tactile, warmth and cold sensations, olfactory, gustatory, pain and pressure, kinesthetic and perceptual spatial orientation), all involve new challenges for construdo to combine the newborn with these stimulations. Destrudo can cause a thrashing of the air with hands, and arms, and feet and legs as the newborn tries to combine and react to its surroundings. At times, construdo disconnects from these stimulations.

There will be alternations in time along the newborn's process. A newborn will sometimes strive to connect with its surroundings, and at other times it will not. These alternations parallel the drives' energy moving along with differentiations between the abs-unc (absolute unconsciou) and the developing DDU, in the three months prior to and following birth.

The Transitional Self/Object Thrasher-Destrudo Trait

In the most turbulent time for the newborn's life (his first six months), its destrudo drive differentiates itself from objects and people. The newborn thrashes its environment or itself while trying to find a target for its anger against the radical change in its living conditions wrought by the event of its birth. In birth, the newborn is thrust into the environment outside the womb. It is unexpected and unanticipated by construdo and therefore construdo is overwhelmed! The newborn has been thrust into a completely different environment, never anticipated or fantasized; and the fact is that its failed construdo arouses its destrudo. But what could be the newborn's destrudo target or object? The newborn has no pre-established response to such unfamiliar conditions, so it thrashes, seeking to combine with an object of its aroused destrudo, either somewhere inside or outside itself.

At these times, newborns have yet to locate or differentiate such spaces or targets. Unable to do so, the newborn's destrudo expressions take on the characteristics of thrashing or flailing the air with wild kicking of feet, swinging of hands and arms. Their torso twists, and the newborn screams and cries. It's the first vocal expressions of anger.

From these earliest expressions of destrudo comes the character trait known as tantrum behavior. Children, adolescents, and adults exhibit tantrum behavior when the frustration of events is so incomprehensible to them they cannot specify what is causing their destrudo reactions, nor do they know specifically what they want to do to rectify the situation. They are unclear about how, toward whom, and for what reasons they desire to focus their destrudo aims.

As unknowing adults, they may exhaust themselves in work or take on the problems of others until they lose themselves completely and become emotionally exhausted. Exhausting themselves emotionally or through activity as a substitute for expression of uncomprehended destrudo aims becomes a discernible trait in these individuals.

The Self-Directed Thrasher-Destrudo Trait

Tantrums directed at oneself reveal attacks from many different directions: "I'm no good. Nobody likes me. I haven't done anything for anybody around me. I haven't done anything I can be proud of." The person may set out to attack an unclear object or the self. These persons have lost a sense of themselves in reality. The result is they feel "beaten up," but are not clear about how or why it happened. This quandary of tantrums and not-knowing refocuses and concentrates only on the self-destrudo from the previous ambiguous stage.

Sublimation of the Thrasher-Destrudo Trait

A sublimation of the thrasher-destrudo trait is seen when individuals over-connect with their jobs and become workaholics. They expend their mental energy in all directions, never stopping to evaluate the appropriateness of their efforts. They just work, work, work... They feel pleased with themselves for having made such terrific efforts. They release thrasher self-destrudo energies in this fashion.

A Reaction Formation against the Thrasher-Destrudo Trait

A reaction formation, or abs-unc (absolute unconscious) reversal of the thrasher-destrudo aim, is seen in people who show the brittle, rigid trait of keeping everything in their lives under tight control. They aim to control the events and people around them. They are not always certain whom they have to control, because this trait stems from a transitional time regarding the aims of self and objects (ended by the sixth month). Heaven help them if they feel they have lost control of their various life situations.

When they can't be in control of the situation, they feel threatened and frustrated, and they may react with some type of emotional destrudo tantrum. They'll inappropriately start yelling, or hang up the phone in the middle of a conversation, or attack the other person on an irrelevant issue, or denounce the person for a character trait unrelated to their shared problematic situation. To the perceptive observer, their emotional stability is precariously balanced.

The Pincer Object-Destrudo Trait

In the original development of pincer destrudo, between six months and one and a half years of age, the infant becomes able to focus visually on objects. Moreover, it muscularly develops a prehensile grasp, so that an object can be taken between the thumb and forefinger. It is the forward destrudo developmental way, carried by construdo, of attacking an object from two opposing directions. It is an advance over the aimless directions of thrasher-destrudo. Thus, the destrudo-arousing object can be caught in a viselike grip and controlled. This control or mastery of the object is the aim of the early pincer-destrudo stage.

Prolongation or continuation of this aim, uninterrupted and unmodified from the DDU (Differentiated Drive Unconscious), leads to the development of the pincer-destrudo character trait of seeking to control everyone with whom one is associated. One's destrudo is aroused by them, because they do the unexpected; this is different from construdo expectations. Beyond developmentally combining with these situations, construdo connects these individuals with a target on the object of their aroused destrudo. These individuals can become "bitchy" when they do not get their way, an expression of the pincer-controlling destrudo character trait. This prolonged, controlling character trait can be sensed if one observes the attitude assumed by these individuals when they employ this destrudo trait. Note how they attack their adversary from opposing directions. For example, "You say I shouldn't interfere with my child's school work. Should I just ignore him?"

The supposed victim position is employed in such a way that causes others to feel guilty. The others concede to what the victim wants them to do, because they don't want to further victimize the supposed victim. The victim unconsciously senses this in the DDU and uses this fact unconsciously to control these others in a pincer fashion.

The Self-Directed Pincer-Destrudo Trait

A character trait from self-directed pincer destrudo is seen in individuals who characteristically put themselves in a dilemma in what they promise to others. The situation becomes a potential attack from one of two different directions. The ones let down will feel object-destrudo toward the person for doing this to them. Putting oneself in such dilemmas reveals a self-pincer destrudo character trait. For example, a man makes a date with one woman for Saturday night, and then absent-mindedly promises to visit a second woman on the same night. One of these women is going to be angry with him when he breaks his date with her.

A Sublimation of the Object-Destrudo Pincer Trait

Sublimation of the object-destrudo pincer trait is seen in people who are characteristically controlling and exacting. They reveal an intellectualizing character constructed by construdo to avoid criticism or intellectual attack from either of two sides, or being caught between two opposing criticisms. For example, these individuals' artistic productions may lack freedom of inner self-expression; their artistic work may be defensive, lacking any reflection of their deepest inner spirit. While there is a sublimation of being controlled regarding freedom in artistic expression, the sublimation is too connected with defensiveness to be a real expression of a free artistic soul.

A Reaction Formation against Self-Pincer Destrudo

A reaction formation against self-pincer destrudo in individuals to put themselves in dilemmas with others is seen when they become predominantly self-focused, not other-focused. Such individuals see only their personal concerns in interactions with others. They talk about their self-interest, ignoring the interest of others with whom they are involved. This looks narcissistic, but in reality they are only reacting against being overly concerned with the interests of others. They will not trap themselves in the middle of concerns, needs, and requirements others have of them. It is a reaction formation against being put in the middle of conflicts of interests between those who surround them and their own self-connections.

The Object Toppler-Destrudo Trait

The infant between one and a half and two and a half years of age wants to move the destrudo object to where it wants it to be in relationship to itself. It is an advance over controlling the object to a fixed position with regard to itself, the pincer aim. The child pulls or pushes the object to where it wants it to be in relationship to itself.

The continued or prolonged aim from this drive stage aim in the adult psyche is the destrudo character trait of wanting to position the opposition to where in space-time one wants it. To do so, a person says things, uses body language, and arranges circumstances that cause one's opposition to move toward to a certain prescribed position, from which the opposition can be toppled. "The rug can be pulled out from beneath them." Developmentally, the destrudo object has come under more influence of the person dealing with it, in construdo-connecting terms.

The Self-Toppler Destrudo Trait

When toppler destrudo is directed against the self over a prolonged period of time, a character trait may develop that causes people to periodically topple themselves. They pull the rug out from beneath themselves, causing reverses in their lives. In business, for example, they play into their competitor's hands, ruin their own business, and go into bankruptcy.

A Sublimation of the Object Toppler-Destrudo Trait

A sublimation of the object toppler-destrudo is seen in iconoclasts, who desire to overthrow some established institution in society in favor of a different political-societal viewpoint. They attempt to change their position from a minority to a majority viewpoint. They would thus topple the existing majority viewpoint.

A Reaction Formation against Toppler Self-Destrudo

A reaction formation against toppler self-destrudo is seen when individuals will not take on competitive toppler destrudo struggles with others. It is because they have an unconscious fear of a tendency in their own DDU (Differentiated Drive Unconscious) to topple themselves. The antagonistic, hostile reaction formation against hostile toppler self-destrudo indicates extreme caution in confronting competitive situations, lest their aroused toppler destrudo turn against themselves. Construdo combines fears of making a toppler competitive error, a misconnection causing these individuals to topple or defeat themselves.

Poking, Preceding the Object Striker-Destrudo Trait

Remember the earlier discussion of *poking*, when a person tries to determine where another person can be struck without resistance, opposition, or retaliation? Whenever this activity goes into character-trait formation through prolongation, individuals show a deep inquisitive nature regarding the vulnerable spots in other peoples' personalities.

The Object Striker-Destrudo Trait

When object striker-destrudo aims enter into character-trait formation, it may be expressed as an abrupt blow at the object of one's destrudo before the other person realizes an attack is being unleashed, or as sudden verbal attacks on another. One might say of an adversary's position on a topic, "That's the silliest thing I've ever heard." The aim is to hurt unexpectedly, when the other person is unprepared for such a verbal attack.

The Self Striker-Destrudo Trait

Turning the striker aim against oneself is seen in the trait of verbally knocking oneself down or hitting oneself in a vulnerable spot. That is, saying things about oneself such as, "I'm too stupid to be a teacher," or "I'm too inexperienced to be a successful actress," or "I could never be a good mother."

A Sublimation of Object Striker-Destrudo

A sublimation of object striker-destrudo is seen in the trait of always building up or promoting perceived strengths in others who would otherwise be competitive. It is expressed in verbalizing perceptions such as "You're a natural businesswoman," or such as "You're a natural athlete," or "Nobody can wear clothes like you!" Keeping the uncertainty principle in mind, there can be other reasons for such a trait.

211

A Reaction Formation against Self Striker-Destrudo

A reaction formation against the trait of self striker-destrudo is seen in the abs-unc (absolute unconscious) reversal and change to another place in the psyche through space-time of promoting one's credentials or promoting oneself. These individuals will say of themselves to themselves, "I'm the greatest," or "I got it all." In behavior, they will demonstrate great conceit and pride. It is a reaction against unconsciously striking themselves.

The Object as Well as Self Smasher-Destrudo Trait

When object smasher-destrudo aims go into the formation of character traits, what is observed is the trait of destroying or wrecking whatever bonds one holds with reality, with others, or with oneself (construdo). In the self-destrudo smasher aim, individuals will wreck or destroy friendship bonds, marriages, or relationships with their children. They may wreck their careers, positions in their communities, or their construdo achievements. All are threatened with annihilation, yet determined by the uncertainty principle.

Such persons may never recover in their own lives from such self-annihilation. On the other hand, they might let their construdo create new lives for themselves. The object of their object smasher-destrudo aim is to smash the opposing object to pieces such that it will no longer exist in reality, physically or mentally. These people walk away from another who is in a heated confrontation with them, but they may never speak to this adversary again. The adversary is mentally smashed out of their personal reality existence. When they do the same thing to themselves, they finish any existence they have in a situation by pulling apart their position in the situation. This trait can be seen when a popular TV minister, who has achieved great popularity with his religious followers, is found to have sinned by stealing funds from the congregation.

A Sublimation of Object Smasher-Destrudo

Sublimation of this trait involves putting things back together for individuals who have broken lives or for themselves when they repeatedly "pick up the pieces and go on." They reclaim a broken marriage or a ruined business with reconstructing construdo. They may do this for others. They may do it in their vocation as a drug rehabilitation counselor, helping addicts who have ruined their lives to rebuild new lives.

A Reaction Formation against Self-Smasher-Destrudo

A reaction formation against the self smasher-destrudo trait involves the more precarious effort to "find oneself in life again." This can happen at certain times when people make self-construdo connections with themselves. It may come at times when they are involved in retreating to self-creative contacts with themselves. This may go back to the aim of the uterine self construdo-creative first period in life, or the self-creative time of childhood from four to six years of age. Clearly, these aims are the opposite of the self-smasher aim. The creative aim here is directed to the self-creative aim by which people remake themselves and become entirely new people.

CHAPTER 18

FORMATION
OF CONSTRUDO CHARACTER TRAITS

This chapter describes how certain early construdo aims continue unchanged into childhood, adolescence, and adulthood. These prolongations or continuations lead to the formation of construdo character traits, as well as reaction formations against them as well as sublimation of them. In the previous chapter and in the case of destrudo character traits, we have already seen how this occurs.

CONSTRUDO TRAITS DERIVED IN UTERO

We human beings have essentially been considered solely as biological organisms. The question becomes: "How is a biological brain that is underdeveloped or rudimentary able to record, remember, or be so influenced by anything from uterine times?" The effects of construdo discussed here reveal its effect on the neural system of gnats, bees, termites, ants – all species that cluster together, stay together, and construct societal organizations. Animals like birds, bats, salmon, caribou, and eels, which migrate to the locality where they were born – all show tendencies to return toward the undermost energy origins of their construdo. These creatures' brains are less developed than that of the human fetus.

It has been assumed that there must be adequate biological cell differentiation, neural brain structure connections, experience in a similar reality to the world, and brain interaction before the fetal brain could produce any influences that would affect the later human organism's behavior. It is assumed that any influence on the fetal brain must be biochemical or originate in the physiology of the brain.

Contrasted with these assumptions is the concept Triadic Drive Theory (TDT) presents throughout this book; that is that he two major drive influences on human beings, construdo and destrudo, do not stem from biological origins or influences at all, but from ever-present forces of physics. These forces influence everything that enters the universe, including a developing embryo and fetus. Therefore, TDT assumes that these constant influences are registered, sensed, and otherwise dealt with in the developing brain of the embryo and fetus.

Freud (1915/1957a) described a drive in precisely such terms as a constant mental influence on individuals of which they were totally unaware. TDT presents destrudo and construdo as the drives produced by just such constant influences.

Construdo represents a mental influence that expands into an individual's world, joins it, constructs in it, and creates one's own world in it; it parallels the energies of the expanding universe. Destrudo, on the other hand represents a mental influence that would break apart all of construdo's work; it parallels the recontracting universe's energies. These mental influences are two aspects of the universal influence for returning whatever has been produced and created by construdo to its undermost energy origins. These mental influences, construdo and destrudo are unifying influences in the psyche.

Origin of the Self-Combining and
Self-Connecting Uterine Construdo Trait

Influences of the self-combining and self-connecting construdo aims are proposed by TDT as being derived from the uniting forces of physics on the developing embryo and fetus during the first six months in utero. Beginning with the fertilized egg, a fused male and female sex cell divides itself into two through the biological process of mitosis, reproducing again into 4, 8, 16, 32, and so on. In TDT terms, as this developmental process occurs, the cells are held together by the uniting forces of physics. In time, the cells begin to differentiate biologically and unite as tissues, organs and organ systems. The developmental progression produces a fetus that has all of the human organs and organ systems of a complete human organism by six months in utero.

The fetus is also held together by the biologically produced connective tissue. These ongoing processes are communicated to the fetal brain through the rudimentary abs-unc (absolute unconscious) – a sensing or registering of a combining and connecting process that has gone on within the fetus's living body. Thus, TDT assumes that as humans become living creatures in the universe, they are influenced by forces that have been constant principles of operation in the entire universe for billions of years.

More on the Self-Combining and
Self-Connecting Uterine Construdo Trait

These two processes, self-combining and self-connecting construdo traits give rise to an early differentiating self-combining and self-connecting construdo energy cluster or fragment in the rudimentary DDU (Differentiated Drive Unconscious) after it has been first registered in the abs-unc (absolute unconscious). This process is the primary sensation of self in every human organism – the beginning of the individual psyche that will connect with itself and enter the world at birth. This differentiating process delivers a sense of what's inside itself to the fetus in the most rudimentary way, in terms of inner thoughts, perceptions, images and feelings. It is the aim individuals revert to when they completely withdraw from reality and go into themselves, the trait of profound withdrawal seen in severely construdo-regressed psychotic conditions. The aim here is to combine and connect with what is experienced in the psyche of themselves without anyoutside stimulation.

A Reaction Formation against Psychotic Withdrawal

A reaction formation against this psychotic trait of deep withdrawal into oneself, or an unconscious reversal of it, is seen in psychotic manic behavior. In these cases individuals in their manic phase act as if they want to combine and connect themselves with everything meaningful in their environment, rather than to withdraw.

Self-Constructing and Self-Creating Uterine Construdo

From six months in utero until birth, the fetus grows steadily in mass. All its structures have developed by the sixth month and continue to grow. No new physiological structures are produced after six months in utero.

Then the first sensing of rudimentary DDU, forming "I," "me," or "self" clusters or construdo energy register a sense of the self as an organism or fundamental psychic energy entity. Uterine construdo either combines and connects with other factors biologically developing in utero or disconnects itself from them. If the latter happens, destrudo fragments of energy oppose the construdo combinations and connections of fragmented or clustered construdo energies in the fetal psyche, differentiating construdo in the rudimentary DDU. These disconnections are also accomplished by clusters of rudimentary destrudo energy that smash apart or demolish parts of what the construdo energies are connecting and constructing. The rudimentary destrudo energy may also topple parts of what these construdo energies have assembled.

As the third trimester concludes and the fundamental differentiation between construdo and destrudo clustered energies is occurring in the rudimentary DDU, destrudo-clustered energies begin to emerge. They break apart the self-connections from the pregnant mother that are initiated by contactual libido progression and the development of the fetus toward birth.

It is proposed that the pregnant mother's construdo-drive attachment toward the fetus affects how the fetus senses itself and its entrance into the world. It may follow that when the mother has mixed feelings toward the fetus, the fetus will have mixed feelings toward itself. Similarly if the pregnant mother has very positive feelings toward the developing fetus within her, after birth the infant, it is proposed, will feel similarly toward itself. The opposite occurs when the pregnant mother harbors very negative feelings toward the infant. It is also hypothesized that this infant will have in its unconscious some very negative feelings toward itself. We see here interaction in the fetal rudimentary psyche of these two drives (construdo and destrudo), as well as genetic forces operating.

The self-construdo creative aim within the self in utero is to break away or disconnect from the host to which it has been connected for a nine-month period in order to produce a new existence as a separate entity. The impending, genetically directed biological birth event also concomitantly fulfills these construdo developmental strivings. Birth accomplishes what the drives of physics seek, although there is no prescribed pattern in such drives for accomplishing this biological progression toward birth. The complementary drive interactions of the two energies achieve independent life and independent mobility. This new life is, nonetheless, physically totally dependent on caregivers for survival. Therein begins psychological connectedness with others and the multiple ramifications for survival.

Birth prompts the capacity to construct and create oneself as desired. Destrudo breaks down or strikes out what is not desired. The two drives show conflicting alternations, which continue throughout life. Construdo groups people together. In these groups there are disagreements and disputes about how things should be done. Destrudo confrontations result. When this conflict is more externalized – from within people to within groups of people – we see wars and conflicts between businesses, cultures, and nations.

In its continuation birth gives rise to prolongations of the rudimentary creative trait. Individuals regress to the continuing creative aim of construdo in their fragmented DDU (Differentiated Drive Unconscious), which can be observed in delusional, illusionary, and hallucinatory systems constructed by some individuals at the psychic level of their rudimentary DDU. Such traits products are the material of psychotic processes existing in conscious psyches.

The constructing, creative prolonged aims will rest timelessly in the fragmented DDU and will be observed in the psychotic trait of building up oneself along whatever false connections these individuals may have made. They create and reconstruct their identities, believing they're Jesus Christ, Abraham Lincoln, Martin Luther King Jr., Teddy Roosevelt, or whoever. The frustration of not enjoying the glory of their constructed identity is explained as a conspiracy against them. These creations and changes in identity are the result of unconscious processes operating on the conscious psyche.

When two subatomic particles collide, a new particle is created; and when a particle moves through time, it becomes its antiparticle – its identity changes. Here and analogy can be made to troubled people connect who impressions from reality that augment or fit into and add to their psychotic systems of thoughts, feelings, and perceptions regarding their false connections with reality.

A Sublimation of the Psychotic Self-Creating, Self-Construction Trait

A sublimation of this psychotic self-creating, self-constructing trait is seen in the creative productions of artists, who, if not for their artistic productions, might become psychotic. Besides this trait they also possess the talent and creative ability to compensate and reground the creative psyche. The mental energy goes into their work, not into creating psychosis. From the descriptions of sublimations in the previous chapter, the reader may infer that these often result in artistic creations. This is true at times, but surely it is not always the case.

A Psychotic Reaction Formation against the Uterine Trait of Self-Combining and Self-Connecting

A psychotic reaction formation against the trait of self-combining and self-connecting can be observed in the trait reflecting psychotic depressions that resist all creativity in a situation or constructing drive productions. This resistance leaves an absence of drive energy for joining the surrounding world in any way, and making nothing of the stimulation from external reality or construdo connecting with it.

CONSTRUDO TRAITS DERIVED AFTER BIRTH

The newborn's construdo, which had previously been self-aimed, on entering the world begins experiencing new types of stimulation. Light, noise, air, body touching, are suddenly imposed on the newborn. Its perceptive, neural-muscular, and kinesthetic systems are experiencing stimulation for the first time in the organism's experience of an external reality.

The newborn's construdo inherently is drawn to combining with the new types of stimulation from the environment. In addition, the newborn continues to combine with itself. Thus, contactual libido from the third trimester finds contact against the newborn's skin in being held against caring adults' warm bodies. This is a transitional time from self-construdo to object-construdo, as well as from self-contactual libido (autoeroticism) toward object-contactual libido, when another person or object becomes its libido aim.

The newborn's construdo automatically combines with the external sources of stimulation and seeks the satisfaction provided by the mother's contact. The mother's prior construdo bonding with the embryo and fetus developing within her uterus establishes an earlier emotional combining bond with the newborn, beyond what the father can achieve. So, too, the mother's contactual libido stimulation of the newborn is more readily accepted by the newborn because of their physical and emotional bond. Hence, the mother will be the first emotional, libido-bonded object, whether the newborn is male or female.

The sections that follow consider self-construdo stages as well as stages of construdo with others, as well as inanimate things. That is, in terms of developmental stages and the aims of these stages. We will not consider all possibilities of these construdo unfoldings, as it would extend the text too far. The concepts will become clear without such elaboration.

The Object-Combining Construdo Trait

The combining aim of the newborn's construdo shifts from self to stimulation, – from the objects in the environment and from the caring significant adults. This shift toward a comparatively more independent life can be observed over the next six months, compared to the completely dependent status of the fetus in utero. In the early object-combining construdo trait, the object is paramount. The newborn's survival depends on it. The newborn disconnects from its self-construdo and mainly combines with the self-caring adults.

The combining with these objects results in a symbiotic idealization of objects. This reality leads newborns to conceive or realize that during this space-time their survival is linked with these caring adults. As a continuing trait there's a tendency to believe that one's survival depends on linking with other persons stronger than oneself. It is one basis for being dependent. By prolongation or continuation of this tendency in the newborn, individuals are often impelled to seek this type of dependent relationship in later adult life.

The newborn's construdo drive automatically seeks combining with the objects or caring, gratifying adults who surround it. They hold the newborn against their warm bodies; they stroke and caress it; they rub their hands over its body; they kiss and nestle their heads against all parts of its body. Here, connecting construdo and contactual libido are acting in tandem.

Further, the infant's contactual needs are transferred to the mouth, lips, and tongue, causing the mouth to be a focus of contactual aims as a new first libido erotogenic zone. This is tantamount to a mouth contactual aim, satisfying two drive aims at the same time, thus making the newborn's satisfaction doubly strong in terms of the two drives (libido and construdo) in tandem. Oral libido dependency and feeding needs become connected with libido and construdo connecting aims. The infant's sucking a nipple, the oral contactual satisfaction derived from construdo and libido aims, and the resultant biological need for nutriment, all solidify the sucking activity of this space-time. But note that there are two drives, not one fused drive as thought in Freud's libido theory.

Intensified by the universe's expanding influence on the abs-unc (absolute unconscious), this object-combining construdo trait can reveal the quality of blind construdo, contactual and oral dependence, libido adherence, attachment, and construdo combining with the object. The object can be an impersonal thing, a territory, or a house as a symbol of a valued connection, just as the significant adults once were to the infant.

The Self-Combining Construdo Trait

When the self is the object aim of an individual's combining trait, individuals will strive to combine with what they sensed or experienced most in themselves. This process involves the person's ego and its perceptions, sensings, and emotions leading to its thoughts. Construdo's developing aims that achieve consciousness are what psychoanalysis has thought of as ego. The drive is more complicated in TDT.

The transitional aims of destrudo here concomitantly cause conflicts with the transitional aims of construdo, due to the alternating energy between the drives and with the drives from self to object. These are the primary drive conflicts in the first six months of life. These conflicts can cause drive-aim confusion during this stage in space-time. The self is its own object of this initial combining stage aim. Its prolongation or continuation creates the self-combining character trait. In this trait people focus on their own perception of the environment, ideation, and emotions – products of their egos.

This is a character trait of borderline and psychotic people. What happens in their surroundings does not stimulate their thinking, reflections, or feelings. This trait reveals people who are concerned or connected only with themselves, despite a stimulating surrounding reality.

This is a phenomenon quite different from narcissism, which Freud (1914/1957c) described as libido taking one's ego as its object. In TDT the self-combining construdo character trait does not involve libido. The conscious ego aspects of narcissism on the other hand are one's personal attributes, looks, appearance, physique, voice, talent, etc.

In people showing the self-combining character trait, the surrounding reality is seen and interpreted only through their eyes. Their viewpoint ignores what others in the shared reality or situation might say is occurring between others and themselves. The extent to which reality is colored by one's personal, self-combined viewpoint will vary with the depth in space-time to which the original self-combining is anchored, namely in utero, infancy, or childhood. The earlier the space-time, the more extensively the personal view of reality will prevail.

When such coloring of reality occurs, these people hold onto their feelings and thoughts about the reality they are dealing with, not caring whether they are right or wrong. They don't care whether they get into difficulties with others around them for holding these viewpoints. The trait causes them to look unrealistic and unduly self-centered, as the larger reality seems to have no impact on them. The trait makes individuals look completely self-determined. Sometimes they appear refreshingly and charmingly childlike; at other times they are maddening to deal with, because people feel they have no impact on them.

Sublimation of the Self-Combining Aim

At a different level of construdo, the self-combining trait can be continued or prolonged into the self-creative construdo aim of seeing reality only from one's own point of view, regardless of what others see, think, or feel. Here there is some overlap with being overly self-connected. Such persons have, however, reached advanced levels of construdo maturity. They may produce works of art that reflect their unique view of aspects of a reality that differ from any view others have held before.

A sublimation of the self-combining aim after birth – an ambivalent trait – is seen in the character trait of being able to compromise two opposing positions. Such a sublimation of the self-combining aim is a noncommittal tendency, a desire to accept positions opposite to one's own, leading to an amalgam of both opposing positions. This ambivalent trait looks for and willingly accepts two sides to every story. It is a precursor, for example, to the readiness to compromise between two differing political positions.

A Reaction Formation against Construdo's Self-Combining

A reaction formation trait against self-combining can be observed in individuals who strive to be overly other-combined or excessively object-combined. Such individuals will combine with others' thoughts, feelings, and beliefs in opposition to their own, disconnecting or mentally blotting out their own thoughts, feelings, beliefs, and attitudes, thus causing them to be overly attached and linked to others. They are symbiotically combined with others, overly dependent on the opinions of others whom they respect or value.

The Object-Connecting Construdo Character Trait

After the construdo self- and object-combining stages comes the construdo-connecting stage, which continues to 1½ years of age. During this stage individuals make, feel, or form connections with the surrounding environment that may be true or false. The construdo focus and striving during this space-time is to constantly connect with reality. Physical and kinesthetic sensations may be one way the child makes such connections.

When connecting-construdo aims are prolonged into childhood, adolescence or adulthood, the result may be the formation of a trait of constantly making connections with what is perceived as reality. Of course, human beings will make connections with their environmental reality that is perceived because of construdo's influence. These connections may be with themselves, and how they perceive themselves, for at this space-time they are also trying to know who and what they are.

Whatever the case, making connections will automatically occur because of construdo's development toward making more and more connections. Thus, the construdo trait of connecting with given realities emerges in whatever way it has been perceived, and continues into later life. When continued or prolonged, the connecting-construdo aim produces the adult trait of making quick connections – true or false connections (even irrelevant connections) – but always making connections.

A Reaction Formation against the Object-Connecting Trait

A reaction formation against the object-connecting trait, or a drive reversal in the DDU (Differentiated Drive Unconscious) – by the reversing change in the TDT (Triadic Drive Theory), in regards to the identity of subatomic particles, is seen in the so-called "free spirited" individuals who are not connected specifically with any of the connections prescribed by society.

The Overly Self-Connected Trait

The overly self-connected trait is seen in those individuals who connect with self-connecting parts of their construdo – in their bodies, ideas, talents, viewpoints, attitudes, or some aspect of themselves they particularly like – and consider these as the only reality that counts. It is the only thing they attend to, look at, or sense as being important – disregarding the outside environment. When continued into later life, a bond is formed that supersedes any connection such persons might have with the outside world.

Such self-connecting parts of one's construdo is proposed by TDT as being an outgrowth or result of fetal sensings of one's mother's rejection or hate of their self-same existence during her pregnancy. In other words, the fetus senses that its survival and its life continuation depends on its own efforts. Survival depends on its efforts alone in the uterine situation. It senses that the pregnant mother cannot be depended on in any way to mentally or physically advance its survival. Thus, from this fetal space-time onward, it is on its own. This orientation continues outside the uterus and is prolonged into childhood, adolescence, and adulthood, and so the individual's life rests solely on what it does for itself. The others around it have no place in its survival picture! Such persons rely only on themselves and whatever survival mechanisms they can devise.

The character trait's aim or direction comes not only from these individuals being overly self-connected but also from their subtle object-destrudo (i.e., pincer-, toppler-, poker-, striker-, and smasher- object destrudo). They do not blame or attack themselves for anything! A psychopathic trait develops in which destrudo catches objects in a pincer attack, overturns them in a toppler attack, finds poker vulnerabilities and exploits them by striker-hitting where it hurts, or smashes the other's psychological existence to make others do as they want. It's all in the nature of hostile, psychopathic, or manipulative type of human connections.

While similarities can be seen between Freud's (1914/1957c) concept of narcissism and TDT's concept of being overly self-connected, these are different in terms of origins. In the former (narcissism), the individual's ego is the object-aim of the individual's libido. In the later (self-connection), and in the TDT language of physics, subatomic particles' operations influence the abs-unc (absolute unconscious) of the confinement property, and cause an intensified construdo connection with oneself. The former (in narcissism) results in self-adulation. The latter (in self-connection) results in trusting and relying on no one but self.

A Sublimation of the Overly Self-Connected Trait

A sublimation of the overly self-connected trait can be seen in adolescent and adult determination to develop a talent or ability individuals perceive in themselves that can invite admiration and social approval. Presumably, such persons will not offend society! They become rock stars, actors, actresses, models, movie stars, dance stars, singing stars, self-centered doctors, great lawyers, outstanding postal workers, special fathers or mothers, and so forth, but in whatever way, special persons in the world!

A Reaction Formation against the Overly Self-Connected Trait

A reaction formation against the overly self-connected character trait is seen in individuals in early adulthood who go through periods of "not being able to find themselves." They head toward one career, then reject it, wondering what they truly want to do. Not until they catch up in time with the unconscious reversals of their goal, if they ever comprehend and understand the turmoil going on within, do they ever settle into some new direction. Generally, they find a new aim for themselves after some tumultuous times in their 20s and early 30s. In short, they are very loosely connected with themselves.

The Object Constructing-Construdo Character Trait

From age one and a half to three and a half years, children construct new combinations of pieces with which they had earlier only been combined and connected. They connect such pieces into larger units or combinations. For example, little dolls are put into a wagon, which the child then pulls behind itself. This action involves construdo connecting the child, the wagon, and the dolls. All advance in a forward direction in space-time.

The child, at this time, constructs relationships with people. According to Ilg and Ames (1955), "He likes to make friends (construct rudimentary friendships) and will often willingly give up a toy or privilege in order to stay in the good graces of some other person – something of which he was incapable earlier" (p. 38).

Putting things together mentally parallels putting interpersonal relationships together at later times. When the object construdo-constructing aim enters into character-trait formation, the trait of putting already existing connections together into larger units of people or things will be observed. These individuals build things from diverse elements or energies that are greater than any of the discrete individual parts. What becomes constructed is a design envisioned by individuals employing this trait. It might be a vision of an organization of people who support a political party or family values, or are against smoking, or support scientists who are against the use of their work for war.

A Reaction against the Object-Constructing Construdo Trait

A reaction formation against the object-constructing construdo trait is seen, for example, in social revolutionaries who want to tear down what has been constructed by society. These iconoclasts strive to destrudo break down (topple and smash) the way people have been put together (arranged) to form a given society in order to construct a very different, new society. Adding this backward-in-time destrudo trend to their forward-in-time construdo ideas, they strive to construct a better society. Because this destrudo-construdo trend is moving in opposite directions, net results are often nil. When applied to an individual, such a process either leads to great success (when the drive motives work in tandem for a new, different self) or to total failure (when the opposing drive directions cancel each other out).

The Self-Constructing Construdo Character Trait

When the self-constructing aim enters into character trait formation, we observe individuals constructing their futures in space-time. By having fantasies of these futures for themselves, they are working at making themselves into the reality of people they want to be like. They visualize, for example, working a certain way in their future career. It is a construdo process to construct step-by-step the planned connections they must make with reality to realize their dreams. They model themselves after the lives of people whose lives reflect what they want their lives to be. They may want to be a banker, a Broadway actor, a congresswoman, a Marine, an electrician, a nurse, and so on. They choose careers (one construdo direction) according to their construdo ego's satisfactions and support of pleasurable fantasies created by themselves in the DDU (Differentiated Drive Unconscious) from earlier times.

A second direction that self-constructing construdo types can construct for themselves comes when destrudo aims merge with the self-constructing aims. One result might be the triumphal fantasy of the self over others, which is created in the DDU, when toppler-object destrudo aims competitively outsmart and outwit others. Such individuals topple others by gaining favorable positions over them. They move others in relation to themselves to where they want the others to be (object-toppler destrudo). These efforts are based on fantasies of being more talented and knowledgeable in their field than adversaries with whom they are connected competitively. In the normal expressions of the self-construdo trait, toppler destrudo is connected to removing whatever obstacles these individuals encounter in themselves.

A Reaction Formation against the Construdo Self-Construction Trait

A reaction formation against the self-constructing construdo character trait is seen in people who are determined to unconsciously fail in their undertakings. Toppler and striker self-destrudo aims connect them with fantasies and fears of their own defeat. Unconsciously – that is, without any awareness – there is a striving to defeat themselves. Their unconscious self-directed toppler and self-striker destrudo aims are directed against themselves and are aimed at pulling themselves down. The unconscious fantasy about their lives, created in their DDU, is that they will not succeed in their life's plan. It is a trait seen in the unconscious notion of themselves as "losers."

Such self-defeating, self-sabotaging attitudes can be the result of an intense sense of guilt that causes their destrudo to turn intensely against themselves. They might have had intensely hostile attitudes toward one or both of their parents earlier in time; now, in an unconscious reversal of their hostility, they believe that they deserve to have such hostility turned onto themselves. Such a defeating self-constructing trait is at times inferred after many observed construdo failed dreams.

The Construdo-Creative Trait

Birth is the first creative experience for all human beings. It is the final rudimentary developmental stage of self-construdo in uterine times. Construdo object-creativeness is seen later, fromthree and a half to six years. It is then that the child imagines a world of purple skies and blue

227

grass; a world in which she is a princess, or is able to fly because of a magic hat; a world in which she can exchange sizes with adults, and on and on.

This is the world children create for themselves, until the surrounding adults point out that it is not actually reality. Their own reality testing tells them their creations differ from reality in many ways. Generally, the children give up their creative way of seeing the world. But if they don't, and this creative character trait continues into adolescence and adulthood, they are able to find ways to reproduce their own original concepts of reality. They do it in the arts, in music, sculpture, paintings, poetry, novels, culinary arts, or floral arrangements. They do it in the sciences when they present original work at the forefront of their field, leading to whole new areas for scientific exploration. It's reflected when mathematicians refer to an "elegant solution." They do it in the business world, when a person like Bill Gates conceives an entirely new business focus, or when Henry Ford first conceived of the production line. This retention of creative ways may be the basis of Freud's regression in the service of the ego.

Love is a Creation of Construdo

The "refinding" of childhood first loves described by Freud is not only Oedipal but also a refinding of childhood and adolescent construdo fantasies of what a loved object might be. Construdo-created love objects have been idealized in such a way that no one in reality could live up to them. The creative construdo of the lovers makes their beloved into what their minds have fantasized in earlier space-time. They are refinding their own earlier DDU (Differentiated Drive Unconscious) fantasies when they find their beloved. Please note that this "refinding" is mostly based on fantasy creations. And that is what love is – namely, a fantasy creation.

A Reaction Formation against the Construdo-Creative Trait

The construdo-creative trait aim is one that everyone experiences at one space-time or another in their lives – for instance, as in being in love, or being born, or creating themselves into desired, planned adults from their adolescence. This construdo-creative trait aim can be reacted against, by energies from the abs-unc (absolute-unconscious) into a reverse direction in the DDU, as observed in individuals who react against what is original or unique in them. Such individuals conceive a creative portrayal of them-selves in society. Yet, they ultimately seek and adopt conformity with others as a safe place to be in society. Being in such a safe place becomes a guiding principle of their behavior.

The Self-Creative Construdo Trait

Some individuals in relation to society show an original and unique flair in personality, dress, or lifestyle. They show the construdo self-creative trait by the way they construct their lives to reflect their overall personal style in relating to their environment. Their lives fulfill early DDU fantasies created in their unconscious during the creative stage of self-creative construdo development (from three and a half to six years), and again later in adolescence, regarding the issue of what kind of adults they want to be. The conception of what kind of person they want to be is revised periodically with new fantasies about how their lives should unfold in their visualized future. It is the construdo drive that creates ideal pictures of what is going to happen in their lives in future space-time.

It should be noted that there can be a factor of object-toppler destrudo here in competition with what those around them think their lives should be. The presence of this object-toppler destrudo is tantamount to an overturning of societal influences on them. It is such object-toppler destrudo that gives them a sense of freedom from the mandates of society. In other words they topple society's influence. There may also be an element of construdo self-created love here for how they have created their lives.

Intertwined between construdo and destrudo aims, libido may take individuals' construdo ego creations and fantasies of themselves as an additional quality of delight and pleasure with themselves. In their sexuality they enjoy what they have created for themselves in the way they participate in sexuality.

Sublimation of the Self-Creative Construdo Trait

A sublimation of the self-creative construdo trait is seen when a parent exerts intense, all-consuming efforts to construct and create a desired type of life for their child. The parent's fantasy is to be lived out vicariously through the child; this child must become someone great. It's a fantasy the parent dismissed or passed over as unattainable in her own life. It comes back into the parent's consciousness as a sublimated fantasy, now wished for in the child.

Reaction Formation against the Self-Creative Trait

A reaction formation against the self-creative construdo trait is seen in individuals who strive to break apart, break down, or knock down a part of their lives. They topple themselves or trap themselves in a self-pincer attack that they consciously and unconsciously set up against themselves. This self-destrudo is the opposite of the positive self-creative construdo trait. This reaction formation leads to a character trait of seeking to be a nonperson – to not know who one is in interpersonal situations and interactions with the world. These individuals lose the boundaries of their own egos. They cannot distinguish what they believe from what others tell them they should believe.

CHAPTER 19

THE DRIVE IDEAL

CONSTRUDO DRIVE IDEAL EXPRESSIONS
AND COMBINATIONS

Sometime before age five and up to age six, and at times beyond, children look at, think about, and reflect on the outcomes of their past object and self-connected construdo behaviors, their object and self-constructed construdo behaviors, and their object and self-created construdo expressions and unexpressed fantasies. So, too, they regard their destrudo and libido actions, feelings and fantasies toward others and themselves in all the different stages of these drives. Such considerations may be conscious, at times partly unconscious, and in other instances completely unconscious. When completely unconscious, the person, although unable to reflect on such considerations, may nevertheless feel or sense something about it.

These self-evaluations and self-judgments stem from actual as well as imagined reactions from parents, siblings, peers, teachers, significant neighbors and relatives, religious leaders, and important loved persons. Children imagine how these figures will react to their behaviors, based on how they understand (correctly or not) the behaviors these figures approve of and disapprove of, how they display ethics and standards in their own social conduct toward others, and what they say about how they regard others. The reactions of these figures, real or imagined, indicate to the children how well or poorly they are relating to these influential figures. The critical factor here is how such events cause the children to feel about themselves, both consciously and unconsciously. Are there feelings of pleasure or displeasure, joy or elation, shame or guilt, self-hatred or self-adulation?

By these ages, children have begun to sense that they are driven by inner mental forces. That is, these mental consequences are the result of the fact that the forces of biology that give rise to libido do not always easily combine with the forces of physics that give rise to construdo and destrudo. These forces are sometimes incompatible and to each other even incomprehensible. Children feel the discrepancies in their drives' strivings. Mentally resolving this incompatibility creates a space-time opportunity for ideals of the construdo and destrudo drives' aims and libido aims. Individuals seek a drive resolution, calming of the drive conflicts and discrepancies. Their minds seek stability and an end to changing impulses.

231

But the drives are often incompatible, because they originate from such different sources in the person: one from biology, the other two from physics. Thus, resolution of drive conflict may not be as simple as the person desires.

Part of construdo disconnects from the rest of construdo and reflects on the value, pleasure, displeasure, emotions, and sensations arising in the child as a result of seeking connections of the three drives, which causes the child to feel congruent within it self when its drives are compatible. Overly self-connected persons have no such concerns. Instead, they continue to focus on how others can be invited to further their own construdo-destrudo concerns for survival: how others have connected with them, pleased them, and furthered their own self-same self-connectedness.

The sensations, the pleasures and displeasures (i.e., good thoughts and feelings versus bad thoughts and feelings about oneself) stem from interactions between the drives. In this sense the discrepancy between those of biological origin versus those whose origin is in physics likely prompts the psyche's need for mediation around these different influences on one's behavior. It is the urge to reconcile the impulses behind the behaviors that have led to varied reactions from people. The disconnected part of construdo with its logical thoughts, feelings, and sensations becomes reconnected in the psyche when these create a new mental organization – the drive ideal.

This drive ideal can come from the self-creative construdo fantasies children have of their future and what they want to be. Often, it's enough to make children feel pleased with themselves. Yet they may not succeed completely in reaching such goals. The degree to which they fail may cause their destrudo to self-attack, consciously or unconsciously.

Later in adulthood, some individuals are prompted to attack themselves in the same way their parents attacked them. They parent themselves through the drive ideal as they were parented, incorporating or copying their parents' reactions. In other instances, copying the parents' reactions can cause the drive ideal to comfort, console, and support these individuals when difficult drive conflicts are faced.

Parents, siblings, significant adults, and peers can voice favorable reactions and approval for these children's mere existence. Such positive reactions can cause these children to feel pleased, good about themselves, and happy to copy the way others react to them.

So begins the early rudimentary developing drive ideal, before three years of age, and further on. It's a space-time in which the child can experience supportive, reinforcing reactions from others. Hopefully these dominate, but not always. Later, between the ages of four and six, children respond to themselves as significant others have. It gives them an emotional

232

and cognitive way of feeling about themselves. They feel okay, good, very good, or even elated. On the other hand, they can incorporate feeling hostile and angry with themselves, condemning, rejecting, and being critical and negative toward themselves.

Self-attack can stem from unconscious construdo self-connecting during drive-ideal formation times. It's unconscious self-destrudo joining the drive ideal. Here individuals are in danger of sabotaging themselves. Unconscious guilt feelings can result in activating unconscious self-destrudo, which seeks to ruin some aspect of their lives. The unconscious guild can be converted into unconscious self-destrudo reactions, as in physical illnesses.

Unconscious attacks on the self are directed by the unconscious part of the drive ideal. Such unconscious attacks have been formed from construdo and destrudo in a manner similar to the way these drives were first differentiated from abs-unc (absolute unconscious) energies in forming the DDU (Differentiated Drive Unconscious). Part of the drive ideal attains consciousness and part does not. It is this unconscious part of the drive ideal in the DDU from construdo and destrudo energies that self-defeating behaviors arise. That these self-defeating behaviors arise is determined by energies that are governed by the replenishment principle.

Present-day emotional conflicts can cause the eruption of emotions from similar conflicts in times past, a return to earlier original energies that were in conflict. Then comes a return to earlier repressed energies, or a return to regressed emotional destrudo times, if these were very significant emotionally. We see self-thrasher destrudo converted into illness when people exhaust themselves in life's activities until they overtax their immune systems.

Without a clear target on which to focus their object-destrudo, they are left at the mercy of the uncertainty principle and the undefined aims of self-thrasher destrudo. According to Triadic Drive Theory (TDT) such ill-defined destrudo aims allow illnesses to progress. Similarly if persons so afflicted had an outward target for their destrudo, as in specific object-destrudo aims, such conversions into physical illnesses would be less likely.

Again, according to TDT unconscious self-pincer destrudo illnesses will occur when people are caught between drive aims; for example, an addictive self-destrudo aim and the self-construdo aims to combine and connect with a drug causing the addiction result in a physically addictive need. At the same time, this physically addictive need will cause a self-destrudo physical illness. Cirrhosis of the liver in alcoholics is one example, emphysema in the lungs in heavy smokers another.

Unconscious self-toppler destrudo may be seen in resistance of the weakened and diminished immune systems to infectious illnesses and resistance to body systems breakdowns, as in backaches, digestive difficulties, or tension headaches. An accidental fracture of a leg, an arm, a wrist, or an ankle may be the result of unconscious self-striker destrudo. A sudden heart attack can fulfill the aim of sudden self-striker destrudo. An unconscious self-smasher destrudo illness may be seen in cancer. In its advanced stages, the cells of the body are invaded and destroyed by the cancer cells.

Self-destrudo can lead to serious illnesses when the unconscious drive ideal's self-punishment aims to weaken the immune system and accept invasion of the body by foreign organisms – bacteria or viruses. This case of serious illness can happen when the unconscious part of the drive ideal turns destrudo against the person. Such a process is serious, because the person is utterly unaware that its own destructive energies are being secretly turned against the self. Consequently, the conscious self erects no defense to its own unconscious self-destrudo. Most commonly, however, an individual's unconscious construdo is connected to advancing its existence in space-time, not destroying it.

A RESTRICTED DRIVE IDEAL

One complexity of the drive ideal's influence can be seen if we consider a few individuals who want to connect with a group similar to themselves, seeking to topple a specific social group. They want to join an antisociety group, a toppler group for another social group. Once united, they progress to exclusively hold destrudo-toppler aims toward the targeted hated group. They can disconnect from their general positive construdo drive-ideal viewpoint, because they are not alone in their destrudo hates. They will not be negatively judged by each other. They now have support in their evaluations of their negative drives' expressions – in their drive ideal. It is based on a new group of people who support them.

This group's shared destrudo may progress to smasher object aim levels. Destrudo's attachment to construdo results in a progression to a more advanced stage of destrudo. Construdo's connectedness even with appalling aims convinces the joiners of the rightness of their cause, as hateful or antisocial as it might be. The Nazi and neo-Nazi movements and the Ku Klux Klan are examples.

THE POWER OF THE DRIVE IDEAL

One uniting aspect of construdo's connections with its own aims (as well as destrudo and libido aims) is found in the powerful, mitigating drive-ideal. Construdo can connect us with society, religion, and community values, as well as with our heroes, parents, teachers, relatives, neighbors, older siblings, and all of their expressed values. The drive-ideal connects us with self-judgments of our effectiveness and ineffectiveness, our pride and shame, and the pleasure and pain resulting from our different behaviors. A strong destrudo disconnecting feeling versus a strong construdo love feeling may result in a couple solidifying their relationship or breaking up. Strong libido versus construdo disconnecting feelings may also be resolved either way.

Toppler destrudo feelings toward authority can cause one member of a couple to break the other's expectations of behavior fitting a relationship. The authority of the expectations is resented and resisted. So, too, it can cause children to break parents' expectations of good behavior from them.

Such conflicts occur between the drives and the drive derivatives in thoughts and affects in the DDU. People can simply mentally retreat from the conflict and ignore it. They may repress the conflict back into the abs-unc, preserving it there mentally, with the repressed idea slipping back into consciousness intermittently when the repressive barrier fails. Such conflict causes the ideas or affects to show themselves as alternates between the two unconscious levels in neurotic and personality-disorder symptoms or behaviors. Lastly, these unresolved conflicts may cause regression to psychotic levels, bringing forth behaviors governed by the abs-unc.

DESTRUDO INFLUENCES ON THE DRIVE-IDEAL

The drive-ideal is a focus of drive energy between the DDU level and consciousness. It allows destrudo aims to enter consciousness, provided these are acceptable to the drive-ideal. The drive-ideal forms between three and six years of age, intensifying in formation by the fifth to sixth year. By then children realize how their construdo, destrudo, and libido aims and expressions have been accepted or rejected by parents, siblings, significant adults, and peers, and, most importantly, by themselves. How similar construdo, destrudo, and libido behaviors are judged in movies, TV, books, the news, in history, in school, and elsewhere – influences children's experiences during the formation of their drive-ideal. From these construdo connections children make judgments about their own construdo, destrudo, and libido behaviors, thoughts, and feelings.

Construdo, destrudo, and libido influences alternate in dominance in the drive-ideal and resulting behavior, due to the universal alternating force. These alternations may be from one drive to another, or to two others. These alternations may last for days, weeks, months, or years. Such transformations of destrudo aims into construdo aims, or into construdo and libido aims, occurs back and forth in the DDU, caused by alternating energies from the abs-unc, which, in turn, influences the DDU. In observable behavior, the drives appear to reverse themselves.

At times individuals express destrudo aims in their behavior. Yet, surprisingly, they have no conscious awareness that they are expressing angry, hostile energies. This is because the destrudo energies from the DDU are unconscious. These people believe they are merely expressing their reactions in how they want to connect with their surroundings. In terms of their drive ideal, they believe and feel they are reflecting society's viewpoint.

Rationalizations of destrudo expressions can be complex and even humorous. Individuals' seven destrudo stage aims promote many different destrudo impulses: these may not always be logical or rational, because they reflect the emotional influences from the DDU on individuals' behaviors.

A mother's defense of her child against those who want to harm it is a different case. The mother's destrudo response is considered understandable, and it is permitted by the drive-ideal as a natural maternal response to the circumstances. The drive-ideal issues no disapproval of the mother. To the contrary, it supports and comforts her.

Destrudo can be aroused in anybody at anyplace and at any time, because it's a natural, human drive in everyone. However, releasing this destrudo in fantasy and imagined vicarious behaviors can be a decision of our drive-ideal in moving away from actual destrudo behavioral solutions to destrudo-arousing human problems. What would we do if carrying out destrudo behaviors were not allowed? What do we do with destrudo connections versus construdo connections when they are in conflict? We may resolve the conflict, retreat from it, or regress to earlier personality solutions.

Remember, from the point of view of physics in Triadic Drive Theory (TDT), the drive-ideal is influenced by the alternating forces of the universe. When the drives operate in tandem, they are in their most powerful combinations. This is true whether the drives are in tandem against others or against themselves. If against themselves, there is an unconscious readiness to weaken the immune system into accepting physical illnesses. If the drives are against others, people can be surprisingly unaware that they are reinforcing unconscious destrudo aims toward these others. By definition of the unconscious, they are unaware of this. They might say, "I like a good fight," where "good" carries the implication of drive-ideal approval. They believetheir cause is just and cannot consider that they are actually expressing wanton unconscious object toppler destrudo.

Destrudo aims in the DDU can be transformed into a righteous drive-ideal cause. The DDU can reverse destrudo aims into a positive conscious-satisfying bonding. This reversal and substitution of energy is due to the replenishment principle of drive energy.

Adolescents, for example, when confronted by an insult, might feel destrudo arousal and an urge to attack the insulter to defend their forming identity. The cause then is in themselves. Their drive-ideal weighs whether a physical attacking response might evoke overwhelming physical retaliation. Such a person might decide angry words are enough, or might decide the insult should be ignored. On the other hand, the adolescent might decide the insult is actually a helpful criticism of a fault that needs correction; that is, thinking of it as a correction can be seen as a help in forming an approvable identity. Thus, the perception is reversed from an insult to a constructive remark.

Self-destrudo allowed by the drive-ideal surfaces when barriers to attacks of others are dropped and vigilance toward worldly damages is suspended temporarily. One might have an accident; e.g., breaking an arm, crashing a car. At certain times, it is acceptable to abs-unc self-destrudo in the drive-ideal that certain self-attack proceeds.

By five to six years (while developing from the third year already), the drive-ideal reaches a level where it is a separate mental unity drawn together by disconnected construdo energies. Thus the drive-ideal has become a separate mental entity that evaluates drive performance, results, and the consequences, which will shape subsequent drive behaviors.

Destrudo energies are subordinated to construdo energies in nature's balance; that is, in the universal energies, as well as in our planetary energies, and in the energy balance within the atoms of our brains. If the balance did not tip toward favoring uniting energies in the universe, nothing would or could long exist.

When the only object drive aim among the drives is a destrudo aim, then the drive balance has gone too far in one direction, beyond its natural position within the psyche: object destrudo is out of balance with object construdo. The drive-ideal must redirect a distribution of self and object construdo versus self and object destrudo, so that it is parallel with the way these energies are distributed in the universe. The optimal existence for any person or all of humankind is one in which mentally there is a clear dominance of construdo-type (joining, uniting, building, creating) energies.

The ideal expression of the two drives is to varying degrees limited by the universal alternating influence from one to the other. Under the uncertainty principle, the alternations may occur daily or weekly in some individuals, for months or years in others, and maybe never in still others. The replenishment principle indicates that the two basic drives' energies can be substituted one for the other in the DDU, but this is almost never recognized by people who are strongly feeling one of these drives. Consequently, they never attempt the substitution. They stay stuck with either drive in excess.

Generally, our behavior is controlled by our drive-ideal. But there are certainly times when the individual's behavior is beyond the dictates of the drive-ideal. We have, for example, seen construdo-created love coupled with a libido attraction and attachment, the drives in tandem, bonding two individuals beyond all conscious choice.

There is a drive-ideal of organizations, businesses, and companies that can be dedicated to achieving stated goals. Some will not be guided by such goals; their drive-ideals are construdo-united in their negation of the organizational goals. Groups can be formed and united in their opposition to ideas of how a group should be functioning or controlled, which the membership rejects in unison. This can also be a destrudo reaction that unites an entire nation. When the group reacts this way toward the actions and intentions of another nation, nations may go to war. It is proposed by TDT that whenever our drive ideal allows the destrudo and construdo drives to unfold very differently from the way their original sources are unfolding in the universe, psychopathology will most likely be the result.

CHAPTER 20

TDT'S CONTRIBUTION TO HUMAN SELF-UNDERSTANDING: 17 ASSUMPTIONS AND 16 SPECULATIONS, A SUMMARY

THE ASSUMPTIONS OF TDT

ASSSUMPTION 1: *TDT assumes the validity of the Big Bang theory of the beginning of the universe.* What is the essence of the Big Bang theory? A super-tightly compacted energy body (about the size of a softball) exploded, releasing matter particles in all directions. The energy was converted into matter, as indicated by Einstein's historic formula, $E = mc^2$.

ASSUMPTION 2: *TDT assumes that Einstein's formula $E = mc^2$ is valid.* The formula stated that energy (E) would be converted into matter particles (m) in an amount multiplied by the speed of light (c), or 186,000 miles per second squared (c^2). The result was a still-expanding universe of matter.

ASSUMPTION 3: *TDT assumes that this beginning explosion was the first breaking-apart energy of physics in the universe.* The energy continues in the multiple destructive energies seen in physics that break matter apart, or strike out parts of matter, or of our mental configurations. The destructive energies might get a grip on an event or thought, such that it cannot be combined with other thoughts.

    The destructive energy thoughts might destroy other thoughts in one's mind completely. These destructive mental processes combine together in the mind by the fifth year to form an unconscious destructive mental drive called destrudo. Destrudo is derived from the destructive component energies of physics drawn from throughout the universe also with an effect on the human brain.

ASSUMPTION 4: *In direct opposition to these breaking-apart energies, TDT assumes that other component energies of physics pulled and held matter particles together.* These are clearly observable in the existing matter bodies of the universe – the forever expanding universe. Such holding-together energies of physics were first seen in the theorized compacted energy particle that exploded into the Big Bang.

    TDT understands that drive energy of construdo originated in the tightly compacted particle exploding in the "Big Bang." Thus, the compacted energy body is seen as the first universal construdo body. This

239

initial explosion catalyzed the universe, one effect of which exploded into construdo bodies having mass, density, and weight. This construdo drive's energies pulled and held together these bodies and particles.

ASSUMPTION 5: *TDT assumes that with respect to ultimate human functioning, these energies of physics produced an unconscious psychic energy of similar character and aim to that of inanimate physical properties – namely to pull mental sensations, experiences, and processes together.* TDT defines this as the construdo drive.

ASSUMPTION 6: *TDT accepts the assumption that after the Big Bang, these energies of physics accidently brought elemental subatomic particles together to form atoms of hydrogen and oxygen.* Billions of years later, these atoms coupled or combined with more complex atoms as carbon and iron to produce living molecules – the forerunner of human life on planet Earth. As Davies (1982) asserted in *The Accidental Universe*, accidents have no intended purpose or reason. Therefore, life's creation on planet Earth was indeed, according to scientific theory, a miraculous, accidental happening.

ASSUMPTION 7: *TDT assumes the validity of Darwinian theory; evolution led to the creation of an unimaginable number of species of animals and plants.* In this respect, life evolved to more and more complex and higher life forms. Human beings are one of the miraculous creations that evolved.

The Mental Characteristics of Humans

ASSUMPTION 8: *Observing how the entire universe holds together, even though parts of it break apart and explode, TDT assumes that these universal forces of physics have had similar influences on the human physical brain (see #5).* The influence is at a very fundamental level.

ASSUMPTION 9: *TDT assumes that this influence of physical forces occurred before conscious thought existed in the human mentality.* These influences were at first unconscious.

ASSUMPTION 10: *Two distinct configurations of influences from universal forces of physics can be discerned.* TDT assumes that one influence could be found in the universe's smallest particles – such as molecules, atoms, subatomic particles – all the way throughout the solar system to galaxies, and onwards to the universe's largest parts, as in

240

clusters of galaxies. In all of these, there existed universal forces that pulled and held the parts of these universal entities together.

Supporting this assertion, are scientifically observed facts. First, the strong nuclear force holds the neutrons and protons together. Second, the weak nuclear force holds the atomic nucleus together. Third, the electromagnetic force holds electrons revolving around the atomic nucleus to the nucleus and holds all things together that exist on planet Earth. Fourth, the gravity force pulls body masses toward each other. Gravity pulls planets toward the sun, but the planets revolve around the sun before they are pulled into the sun's gravity force. Gravity pulls stars and planets in galaxies toward each other and holds the galaxies together. In billion-year cycles gravity pulls galaxies toward each other; before, alternating, they move apart again from each other.

ASSUMPTION 11: *TDT assumes that these four forces (strong nuclear force, the weak nuclear force, the electromagnetic force, and gravity) affect the atoms of the brain.* These forces cause the formation of a psychic mental-drive energy, before conscious mental thinking and rationality existed. They process stimulation to the brain and cause the stimulation to be joined and connected together into sequential stimulation and response (S-R). In the still unconscious psyche, sensations, visual images, feelings or emotions, flitting thoughts, and ideas were then united. TDT identifies the psychic energy causing such combining and connecting of such stimulation and resulting response-construdo energies or the construdo drive.

ASSUMPTION 12: *TDT assumes that an opposite psychic-drive energy came into universal existence.* This opposite drive energy sought to break apart and break down whatever the construdo drive energy had put together. The destructive energies of the destructive drive are constantly augmented and reinforced by energies from the multiple forces causing explosions and destructiveness, as well as damage to existing matter bodies. TDT calls this destructive-drive energy the destrudodrive.

The Libido Drive

ASSUMPTION 13: *TDT assumes that Freud's biologic drive – libido, the human sensual, sexual drive – operates on the human mentality in conjunction with the two drives, construdo and destrudo, derived from physics.* Remember, Freud conceived and invented drive theory, his "dual drive theory," to which he later added the aggressive drive to the libido drive (Freud, 1930/1961a). Freud (in the time pre-dating modern physics) thought of human beings as only biological creatures. Hence, he thought the aggressive drive was naturally a biologic drive. However, biology has never confirmed the biological origins of the aggressive drive. To this point TDT has incorporated Freud's libidinal drive into its theory of three psychic drives – hence, Triadic Drive Theory (TDT). Thus, Freud's biologic drive, libido, was retained in TDT, but his biologic aggressive drive was not. TDT regards the aggressive drive as derived from forces of universal physics, and not derived from biology.

ASSUMPTION 14: *What has been asserted by TDT leads to certain philosophic conclusions about the nature of human life on planet Earth, as well as possible life elsewhere in the universe.* TDT rejects the assumption that the life on planet Earth is the only life in the universe, and that life on planet Earth is the only type of life that can exist in the universe. Living creatures in the universe might come from different, unknown existences of energies that are outside the conception of humans on planet Earth.

ASSUMPTION 15: *Because TDT sees humans as integral parts of the universe's energies, it assumes with high probability that humans as creatures created by the universe will find new horizons for potentialities in their universal thinking.*

ASSUMPTION 16: *TDT assumes that the unconscious psyche is the source of an unconscious mentality, and that both exist.* Other disciplines of psychotherapy, such as cognitive therapy, question these concepts and rely on only what can be seen or cognitively observed. These unconscious concepts do not fit their criteria, or, are even outside of their parameters.

ASSUMPTION 17: *TDT assumes the validity and accuracy of Freud's formulations about the workings of the unconscious mind.* In *The Interpretation of Dreams*, Freud (1900/1953a) offered a conceptualization of how the unconscious works, based on observations of his own dreams. He extended his conceptualizations to the dreams of others as well. Freud posited that the workings of the unconscious were reflected in dreams. However, he did not distinguish between a construdo psyche and an unconscious mentality, as has TDT.

THE SPECULATIONS OF TDT

TDT and Religion

SPECULATION 1: *The building up, uniting, positive, constructing, creating energies of construdo, versus the breaking-down/ breaking-apart negative energies of destrudo call to mind the Good-versus-Evil dichotomy of the Christian religion – the personified concepts of God and the Devil.* TDT's theories, however, are derived from scientifically observable facts. Triadic theory has made inferences from the facts about the nature and functioning of the human mentality. This overall structure developed from Triadic Drive Theory (TDT) has been extended to include all living things.

Religion rests on faith in the truth of the word of God, as reported in the Bible. These words are not established scientific observations. Rather, these words and/or concepts are reported as inspired by God and delivered through various human sources. Nonetheless, both TDT and the words of the Bible arrive at similar conclusions about the nature, characteristics, and striving tendencies of the human mind. The constructive, positive striving, building-up, achieving energies of the construdo drive are comparable to the good and good-striving sides of human beings presented in the Bible, as in to follow the inspiration of God and Jesus as he conducted his life to reach these achievements. The converse striving or breaking down of whatever construdo has put together is the destrudo drive. It would break apart and break down construdo achievements. In the Bible this breaking apart or down is personified as the Devil.

Philosophy and TDT

SPECULATION 2: *TDT speculates that if other creatures exist in the universe, then they have developed and perfected their construdo drive far beyond the level that humans have here on Earth.*The concept of living creatures in spaceships capable of reaching planet Earth from elsewhere in the universe carries definite implications for TDT. The creators of such spaceships would have extensively developed their construdo drive energies far beyond humans on earth to validate such an achievement.

TDT asserts that the construdo drive (not destrudo or libido) is the primary psychic energy that will further human continuance and advancement in the universe. Humans are creatures of the universe's creation. Humans have far more potential for understanding the universal forces that made them. Further, humans can and use these forces for new and never-conceived accomplishments in the universe.TDT reveals such possibility.

Other Universal Forces on the Human Mentality

Some of the universe's forces of physics do not directly influence specific aspects of the human mentality. Rather, the influence can be on overall ways the three drives solely operate and how they interact.

Humans often display alternation or oscillation in their emotional feelings toward others and things, as well as in their drives, to construdo connect and attach themselves to others and things. They can move from construdo love to destrudo hate. They can connect with a political movement or a religion; then, they disown it. They adore a marriage partner; then, they perhaps even come to hate that same partner. They crave living in certain environments; then, at some point, they lose that passion. They can extol a certain type of music, literature, or cooking, but these tastes can change. They can champion science – then a short time later, they may even turn against everything scientific.

TDT traces these reversing emotional and drive attachments to the alternating quality in many of the universe's energies. Such energies influence the human unconscious psyche, resulting in alternations in human emotions and drive directions, from construdo to destrudo. Moreover, these alternations motivate the human tendency to reveal behavioral cycles. The planet shows periods of rain, alternated with drought. A man feels libido drive, followed by periods of abstinence. Humans can be very productive, and then feel lethargic. Humans can be fun loving, then calm down and want to live a serious life.

Alternations in the universe's energies are seen when galaxies pull together, and then move apart, in millions-of-years cycles. In the solar system, sunspots appear, then gradually lessen, and disappear, then reappear in eleven-year (or longer) cycle alternations. The magnetic poles of earth alternate in cycles over millions of years. The North Pole becomes the South Pole, and the South Pole becomes the North Pole.

Again, at the subatomic level alternations are seen in the atoms of the universe. Subatomic particles smash into each other, destroy each other, and then form new particles, continuously, within the universe.

SPECULATION 3: *TDT speculates that the String Theory is an alternation of particle physics taken to an advanced level.* TDT believes these two levels, namely, psrticle physics, a first-level concept of reality; and the string theory, a second-level concept of reality – are two distinctly different levels of reality. It should be understood that TDT is based on the first level of reality, particle physics.

SPECULATION 4: *TDT speculates that these alternations of the forces of physics occur at all levels in the universe's existence, and is the source of humans' need for and acceptance of change in their lives.* This speculation is an instance of TDT's concept of psycho-physics, in which the universal forces of physics have an overall general influence on the behavior of humans and on the human mentality. First, an influence is observed on the human mentality; second, a force of universal physics is deduced that could have caused the influence.

The latter situation that physics has an influence on human mentality is observed when a person responds to a present-day emotional situation as if it were actually a similar emotional situation in the person's past. Such a situation is seen when a father dominated and controlled every situation in a child's development. This child may grow up to be an adult who resents directions or orders from a boss. Such a response is inappropriate in the reality of the later job situation. The inappropriateness of the response calls attention to possible psychological influences from elsewhere in this adult's life

Reactions from this individual's past have determined reactions to the present-day situation. Such reactions Freud called instances of the repetition-compulsion a compulsion to repeat events that were most painful in the individual's earlier life.

The energies of the psyche causing such out-of-time reactions, TDT asserts, are determined by energies of physics to return to undermost old energy sources of physics, in response to a given new-stimulus energy situation. Such responses are set off by emotional similarities between the present situation and old similar emotional configurations from the past situation. The emotional similarity is the key to reviving the old reaction. Returning to old responses, instead of responding to the present situation, stems from the underlying principle of physics governing the situation, namely to return to undermost energy origins.

SPECULATION 5: *TDT speculates that forces of physics caused the mental happening observed in giving an old, out-of-time response, instead of a response to the present stimulus situation.*The tendency in forces of physics to return to their undermost energy origins is a force yet to be found or documented in physics. Yet, TDT believes it will be found and corroborated.

The world-renowned physicist Stephen Hawking (1988, 2001) advanced the concept that the universe would go from Big Bang to Big

Crunch, in which forces of gravity would cause the universe to contract into itself. The result would be another collosal Big Bang explosion. It is thought by TDT that such a Big Bang could start the beginning of another universe. If Hawking's conception is true or valid, then TDT's concept of the universal energies of physics returning to their undermost energy origins would be confirmed. His concept of the universe reverting to a Big Crunch indicates that the universe (as we know it) will ultimately end.

SPECULATION 6: *TDT speculates that humans' lives come to an end because humans' lives are commensurate with the universe's existence.* Remember, humans are an integral part of the universe. Therefore, humans will die because the universe will die.

The Part Reflects the Destiny of the Whole

SPECULATION 7: *TDT speculates that what occurs to part of a system may reflect what will ultimately occur to the entire system in some fashion.* If TDT's concept of returning to undermost energy origins applies to both universal psychological happenings and universal happenings of physics, and if the forces of physics follow this same principle, then what happens with respect to the exponential gravitational pull of black holes reflects what will in time happen to the entire universe; that is, something akin to Hawking's Big Crunch The time may be another 13.8 billion years, as long as it has taken to reach the universe's present existence.

TDT has presented a new way to study the relation of psychology to physics – by looking at the psychological functioning of humans and then inferring what kind of forces of physics could cause such psychological functioning. This could lead to discovering heretofore unseen forces of the relation of psychology to physics. Such questioning also can potentially deepen the study of psycho-physics.

Depressive Mentality Caused by Universal Forces of Physics

SPECULATION 8: *TDT speculates that this universal "contracting" tendency can influence human mentality, causing emotions to contract back from the surrounding world.* This potential contracting tendency may cause depressive feelings in one's unconscious psyche. The universal contracting influence can become imbalanced with feelings that individuals sense regarding ongoing expanding life ahead for these individuals. In instances of extreme influences, the unconscious psyche with theoretically sense an end of the person, or sense a developing death expectation.

246

SPECULATION 9: *In the unconscious psyche, this expectation of "the end" becomes a goal to take control of and to direct.* Initially, this unconscious expectation of "the end" is an unconscious reversal of the wish to live. Basically, it all translates into a psychotic depression as in seeking to commit suicide. These depressive emotions are expressions of striker and smasher destrudo aims directed at oneself.

SPECULTION 10: *TDT speculates that when the psyche senses that humans exist in an expanding universe, the universe's actual energies create emotions of positiveness and optimism.* These emotions constitute the source of hope for better times or in TDT terms, for positive construdo connections and creations.

SPECULATION 11: *TDT speculates that this universal expansion translates emotionally to positive joining and connecting things in individuals' lives.* The expanding universe gives rise to, in a word, construdo. From these positive, building-up, construdo energies, humans draw the positive anticipating mental energies that carry them forward to many positive expectations and happenings.

SPECULATION 12: *Should the positive connecting construdo drive become imbalanced with regard to the other drives, drive derivatives, and emotions, the result is likely to be a psychotic manic state.* In such cases these individuals expect or anticipate too much from more and more intensified construdo connections. In lesser instances of such anticipations, emotional incongruity and disappointment occurs. But in full bloom, intensified construdo out of balance with all else in the mentality of these individuals creates some variety of manic euphoria. Beyond this level of construdo intensification, still greater levels of psychotic manic states begin to occur, as in bipolar conditions. In other words, too much construdo energies flowing into individuals' lives at a given time do not lead to improved positive mental functioning.

Creation of the Unconscious Psyche by the Universal Forces of Physics

Freud (1900/1953a) found that a dream was a series of visual pictures for the dreamer, not a series of ideas. However, thoughts, ideas, and emotions were associated with the visual images of a dreamer's dream. Often these thoughts, ideas and emotions could be ascertained if the dreamer said what came to mind when thinking freely about a visual image from the dream. It's what Freud called "associations" to the dream images.

Freud found that the visual images of a dream were often condensations of several ideas connected together into one visual image. The ideas were "fused," transformed, or "condensed" into the dream's visual images. Sometimes such fusion might follow principles of symbolization, where a given image was an artistic or poetic expression of several ideas. All of these fusing dream processes Freud called "condensations." Dreams also revealed a moving of intense feelings from one emotionally charged site in a dream to a lesser charged, or more neutral, emotional site. Freud called this dream process displacement.

As referred to earlier, Freud found that dreams do not follow logical time sequence. Dreams are timeless. What is remembered as the beginning of a dream may actually have occurred at the end of the dream experience, and vice versa.

When we recall a dream, nonetheless, we organize it in our conscious, logical reporting of the sequence of the dream's visual images. Freud called this report the "manifest" conscious recounting of the dream. The condensed, displaced, timeless actual images of the dream Freud called the "latent" dream content. It is the latent dream content that we recast in reporting the manifest dream. We make the report logical, comprehensible, and meaningful in our consciously arranged report. The manifest dream content is what we consciously reconstruct and remember. It is the descriptive dream "from above."

Freud found that the unconscious operations were neither logical nor rational. He found, however, that there was a systematic codified way that the unconscious operated. He detailed the principles of the unconscious system of operation in *The Interpretation of Dreams* (Freud, 1900/1953a). TDT concepts of the unconscious are based on Freud's formulations of the unconscious.

TDT and the Unconscious

TDT distinguishes two levels of the unconscious. The first level involves those fundamental mental processes that regulate and direct the human body's living processes. Humans do not consciously think about these processes to keep them living or operating. Examples are: breathing, heart beating, waste elimination, blood flow, reactions to heat and cold. These are established in utero and shortly after birth. These unconscious influences TDT calls the "absolute unconscious" (abs-unc).

SPECULATION 13: ***TDT speculates that as the mental-drive energies differentiate (while keeping a balance), the second unconscious level crystallizes.*** These differentiations of the mental drive energies form the drive differentiated unconscious (DDU). Such differentiations occur during the second and third trimester in utero. Further mental drive energy differentiations occur following birth, and continue into the fifth and sixth year of life. If these drive energies become radically out of balance, then mental psychopathology results. The earlier in life the drives become unbalanced, the more severe will be the depth of psychopathology.

TDT understands that uterine imbalances produce psychotic conditions or severe, pervasive, all-encompassing psychopathology in the individual's orientation towards the environment. If construdo is under-developed in relation to the other two drives (destrudo and libido), this lack of development of construdo can in TDT thinking produce an autistic mental condition. If construdo does not keep the abs-unc (absolute unconscious) separated from the DDU (Differentiated Drive Unconscious), the two levels of the unconscious may alternate between each other while moving into consciousness, thereby producing schizophrenic pathology.

Early after birth, imbalances cause disruptions in aspects of the personality, producing personality pattern disturbances. Such disturbance occurs before the infant child has had complicated interactions with others in their environment. The borderline personality is a case in point. Here, imbalance occurs while the infant is trying to distinguish feelings of self from others. Is it the feelings of oneself or of the mother? The infant is not certain. If the imbalance occurs at this time, it may remain a problem for life.

Later in childhood, when individuals have progressed in the development and differentiation of their drives, an imbalance can occur between their drive aims and wishes.In such cases the environment will judge their wishes to be unacceptable, and will, socially, reject such individuals for expressing such wishes. The destrudo striker desire toward one's father, for example, may be unacceptable and threaten ostracism in the family if the desire persists. This kind of conflict may produce neurotic psychopathology. If the drive desire becomes unconscious but persists in the unconscious, the conflict then exists between conscious behavior and unconscious desire.

Operations of Subatomic Particles and the Operations of the Unconscious

Triadic Drive Theory (TDT) based in physics utilized the uncertainty principle. As has been discussed earlier, when a particle's position could be determined, its speed could not be determined, and vice versa. How particles would act could only be stated in terms of probabilities. A particle's identity would depend on its position in space-time. Predictions about particles' future actions could not be made. Rather, particles' actions could only be explained after the fact. Such physics did not deal with the reality of our everyday world. This physics was above and beyond cause-and-effect concepts of Newtonian physics. Similarly, it was noted that Freud's principles of the operations of the unconscious did not follow, nor were they based on, the logical, rational reality of cause-and-effect concepts of Newtonian physics. For example, an idea can underlie two successive, or different, visual images in a dream, yet still be the same idea in Freud's concept of the unconscious; this is just as a subatomic particle can be at two different places in an atom at the same time. This conceptualization is beyond human common or simple logic. The changing of the visual image of one idea is the dream's unconscious concept of displacement. The idea is moved, or displaced, to another place. The same idea can be represented at several different places in one dream image. Time does not exist or apply in the unconscious, nor does it exist in the subatomic particles within the atom. Both are in a different reality than that of the consciously known world of logical forward moving time around us.

Another operation of the unconscious of dreams is seen in the unconscious tendency to pull together several separate ideas into one visual dream image representing all of the ideas. For example, several people can be fused into one dream image of one person.

One visual dream image can be a symbol for several ideas or concepts. This is the case in Freud's "symbolization" process of gathering together dream ideas. The dream image of a window with bars, for example, can symbolize jail or prison, and represent dreamers' feelings of being trapped by their life situations and feelings. They may see no way out.

SPECULATION 14: *TDT speculates that what Freud found in the "condensation" process of the unconscious is derived from the "confinement" process of subatomic physics to reduce its elements into a smaller and smaller particle.* Reversing of an idea underlying a dream image was another process Freud found in the unconscious productions giving rise to dream images. The idea condensed and underlying the visual dream image is the reverse of what is seen in the dream image. Stepping up or having a elevated sense in a dream situation may portray actual feelings of being let down in the situation by others, or of not achieving what was originally sought. With respect to physics as the template for the appearance of psychological phenomena, the unconscious feeling and idea reversals, seen in dreams, is a reflection of the subatomic process that continually changes subatomic particles into their opposites in the atom. Thus all of the subatomic-particle operations do not follow any time sequence, nor do they follow any time sequence in the human unconscious. Influences from subatomic particles come in and out of the subatomic particle existence without reason or influence from the human unconscious.

A new unconscious influence is now seen in the unconscious. The lack of any logical reason or rational reference underlying such occurrences portrays the nature, or essence of the unconscious mind. Again, there is no sense or logic behind the operations of the unconscious mind. The subatomic particle's influence on the unconscious mind and Freudian principles of unconscious operations show remarkable similarity to one another. Those similarities of the physical and psychological are both seemingly irrational yet systematic in their particular operations.

The Developing Unconscious Psyche

In the developing mentality of the growing infant, visual images, sensory sensations, fleeting ideas and feelings, or emotions combine together in the brain's first developing capacity to collect or hold together these mental occurrences. Such occurrences crystallize before these occurrences have gained any awareness in consciousness. The physical brain structure produces this functioning for the first time. TDT labels it the unconscious psyche. All of the aforementioned mental occurrences are being retained, for the first time, in the physical brain. But this retention occurs before there is any awareness of what is being retained. Gradually, in time, such retention begins to show some effect on the infant's awareness. The infant learns that crying can cause results in its environment. The mother may pick up the crying infant, rock it, or comfort it in some way. The infant makes the connection in its mind. This connection is early human learning, the beginning of the human mentality. It has progressed from the brain, a physical organ, to a conscious mentality of the infant's awareness of itself in a rudimentary way. During these times and beyond, the drives are entering the infant's consciousness, and beginning to direct the infant's behavior.

REGULATING THE HUMAN DRIVES: THE DRIVE IDEAL

As individuals develop from infancy to childhood, they may experience conflicting feelings from their construdo drive energies versus their destrudo drive energies even towards the same person; and, still more confusing, the impulses of these two drives (derived from forces of physics, versus the libido drive), also derived from biological energies. Feelings of alienation or estranged mentality can be the result of such drive conflicts. Seeking to be conflict-free, individuals at these times begin control their drive impulse. Gradually, they sense, feel, and think of controlling their drives mentally by some inner part of themselves. Individuals vaguely sense a part of themselves that might control the interactions of their three drives' with each other. In TDT terminology, this is a part of the construdo drive, the "drive ideal," reflecting the way their drives have worked best together in the past and want to work similarly best in their future.

The judgment of what drive expressions worked best is based on these individuals' experience with various expressions of their drives in the world around them: which drive expressions have led to praise or approval from others around them, which have led to condemnation or criticism, or have led to moderate reactions, both positive and negative. The result is a mental agency that attempts to govern the construdo experiences based on construdo's positive success in reality versus construdo's negative results in reality.

By the time each of the three drives have come to full development (at five to six years of age), humans have experienced many reactions to their drives' expressions – from parents, relatives, friends, peers, teachers, authority figures, group members, church members, community block organizations, all influential overseers. Thus, the drives become tempered by reality influences surrounding the expression of the drives in human beings.

SPECULATION 15: *TDT speculates that from these comparable experiences, humans have developed the drive-ideal so as to bring each drive to where it is functioning at its best balanced level with the other drives.* Observation, although not based on experimental investigation, suggests that construdo works best at the connecting, constructing level. For construdo to work less well would not be enough for the drives to have positive influence on each other. For destrudo, an optimal level is at the toppler and striker levels. The thrasher and pincer levels are not sufficient to deal with many reality situations, while the smasher level may be too threatening or frightening for many situations.

Mentally healthy existence in families, groups, communities, societies, nations, and civilizations depend on the continuation of the drive ideal. It has also influenced the laws governing these communities, societies, nations, and civilizations. This influence has mainly come from religions. Buddhism, Christianity, Islam, and Judaism have carried their influence into the societies and nations in which they predominate. Without the drive ideal in some form (primarily religion), humans would experience chaos in their drive's expression. The drive ideal's control of radical expressions of the three drives keeps them in a healthy drive balance. TDT calls it "being of good conscience." Striker-destrudo energies may still be employed by the drive ideal to knock out undesired drive impulses from the unconscious and conscious mentality. Bad consequences can often follow in situations when striker and smasher destrudo urges are carried out by large groups of people, convinced of the "rightness" or "goodness" of their cause. This follows from the overall principle of TDT that keeping destrudo expression to a minimum in most human interactions is the better part of wisdom.

SPECULATION 16: *TDT certainly does not believe that all destrudo expressions should be eradicated.* Some causes are just, as when they oppose bad construdo alliances; for example, the Nazi movement, or out and out slavery – assuming that these value judgments can be defined, in time, as valuable for positive furthering of the human species.

In presenting the revolutionary concept that humans are mentally an integral part of the universe, TDT can perhaps open new horizons for philosophic speculation as to what it is to be human and as to what to be a part of the infinitely vast universe. Just as thinkers have for ages theorized the meaning of life on Earth and being human, now theorists can begin to speculate on the meaning of being a mental part of the universe. It opens the case for new and different horizons as to where human thinking came from and where it can lead us.

Being more than biological creatures is certainly hard to contemplate. Human beings have always thought about what was passed on to them biologically from earlier generations, as well as taught to them by earlier generations. Now human beings might begin to think about what is being passed on to them from the universal forces of physics.

Moreover, human beings can begin to think of what they can learn about forces of universal physics that have at least theoretically caused certain phenomena in them. This is new way to study physics and a new way for human beings to study the psychology of the self.

Humans might carefully consider the influences of universal forces that have not always led them into the best directions for enjoying life. For example, the universe's alternating energy produces a desire for changes that may not be needed. Individuals might reexamine tendencies to contract withdraw into themselves.

Humans might also learn to be careful of inner tendencies to magnify destrudo's universal influence on themselves. Such a tendency can cause a focus on others as enemies. Humans, for destrudo reasons alone, without construdo evaluations of the situation can quitea often be too quick to kill their enemies. In the same manner, they too quickly go to war.

In constructing this Triadic Drive Theory (TDT), my formulation of construdo drive has hopefully enabled a door to open so that a vast amount of data can be brought to bear on the template that relates the psychology of the person to basic properties of the world of physics.

AFTERWORD

Dr. William Johnson has given us a tour de force — a bona fide original conception taking Freud's psychoanalysis to a profound, daring, and significant step further. This Johnson's work, *Triadic Drive Theory*, is such an example of what happens when a scientist/clinician opens our eyes to a new and perhaps fantastic idea.

Copernicus did it in his heliocentric discovery, by demonstrating that it is the Earth that revolves around the sun, and not the other way around. Darwin did it by demonstrating that evolution 'evolves,' so that it is characterized by a past, a present, and a future, and therefore that Homo sapiens has a relation even to the amoeba!

Freud also gave us this kind of challenging gift, in a nutshell telling us that we may not even know our own minds, and therefore that the meta-psychology of psychoanalysis can give us access to such mentation and to all of its vicissitudes.

Similarly, in an important way, Johnson challenges us to see that Freud, in fact, was right to unfold psychoanalysis into biology; but perhaps psychoanalysis must by all means be taken even more fundamentally — down to physics! And that is what Johnson has done here. He has not only opened the door to suggest the connection of psychoanalysis to physics, but has spelled out this connection highly specifically, by offering us the "updated" drive theory.

After reading this work, one cannot come away without feeling a cluster of emotions and characterizations. These include characterizations such as "fascinating," "audacious," "unprecedented," "courageous," "ground-breaking," and yes, along with all of this, a final epiphany revealed about Dr. William Johnson himself (in his "mad" pursuit for scientific truth) — a piece of genius.

Henry Kellerman, Ph.D., ABPP, psychologist, psychoanalyst, author.
Dr. Kellerman's latest books include *Ghosts of Dreams* (Barricade Books) and *There's No Handle On My Door: Stories of Patients in Mental Hospitals* (American Mental Health Foundation Books).

REFERENCES

Adler, A. (1956). *The individual psychology of Alfred Adler*. New York, NY: Harper.

Asch, S. E. (1952). *Social psychology*. Englewood Cliffs, NJ: Prentice-Hall.

Brenner, C. (1971). The psychoanalytic concept of aggression. *International Journal of Psychoanalysis, 52*, 137–144.

Capra, F. (1975). *The Tao of physics*. Berkeley, CA: Shambala.

Chester, M. (1978). *Particles: An introduction to particle physics*. New York, NY: New American Library.

Darwin, C. (1972). *On the origin of species by means of natural selection, or the preservation of favoured races in the struggle for life*. New York, NY: AMS Press. (Original work published 1859)

Davies, P. C. W. (1978). *The runaway universe*. New York, NY: Harper & Row.

Davies, P. C. W. (1982). *The accidental universe*. Cambridge, England: Cambridge University Press.

Dollard, J., Doob, L., Miller, N., Mowrer, O., & Sears, R. (1939). *Frustration and aggression*. New Haven, CT: Yale University Press.

Fenichel, O. (1945). *The psychoanalytic theory of neurosis*. New York, NY: Norton.

Freud, A. (1936). *Ego and the mechanisms of defense*. New York, NY: International Universities Press.

Freud, A. (1946). *The psycho-analytical treatment of children*. New York, NY: International Universities Press.

Freud, A. (1972). Comments on aggression. *International Journal of Psychoanalysis, 53*, 163–171.

Freud, S. (1953a). The interpretation of dreams. In J. Strachey (Ed. & Trans.), *The standard edition of the complete psychological works of Sigmund Freud (SE), 4*. London, England: Hogarth. (Original work published 1900)

Freud, S. (1953b). Three essays on the theory of sexuality. *SE, 7*, 123–246. (Original work published 1905)

Freud, S. (1955a). Beyond the pleasure principle. *SE, 18*, 7–64. (Original work published 1920)

Freud, S. (1955b). On transformations of instinct as exemplified in anal eroticism. *SE, 17*, 127–133. (Original work published 1917)

Freud, S. (1955c). Studies on hysteria. *SE, 2*. (Original work published 1893-1895)

Freud, S. (1957a). Instincts and their vicissitudes. *SE, 14*, 109–140. (Original work published 1915)

Freud, S. (1957b). The unconscious. *SE, 14*, 159–209. (Original work published 1915)

Freud, S. (1957c). On narcissism: An introduction. *SE, 14*, 67–102. (Original work published 1914)

Freud, S. (1957d). On the universal tendency to debasement in the sphere of love. *SE, 11*, 177–190. (Original work published 1912)

Freud, S. (1958a). The dynamics of transference. *SE, 12*, 97–108. (Original work published 1912)

Freud, S. (1958b). Formulations on the two principles of mental functioning. *SE, 12*, 213–226. (Original work published 1911)

Freud, S. (1959a). Character and anal eroticism. *SE, 9*, 168–175. (Original work published 1908)

Freud, S. (1959b). Inhibitions, symptoms and anxiety. *SE, 20*, 77–178. (Original work published 1926)

Freud, S. (1959c). The sexual enlightenment of children. *SE, 9*, 129–140. (Original work published 1907)

Freud, S. (1960). The psychopathology of everyday life. In J. Strachey (Ed. & Trans.), *SE, 6*. London, England: Hogarth. (Original work published 1901)

Freud, S. (1961a). Civilization and its discontents. In J. Strachey (Ed. & Trans.), *SE, 21*. London, England: Hogarth. (Original work published 1930)

Freud, S. (1961b). The economic problem of masochism. *SE, 19*, 159–172. (Original work published 1924)

Freud, S. (1961c). The ego and the id. *SE, 19*, 3–68). (Original work published 1923)

Freud, S. (1962). The neuro-psychoses of defence. *SE, 3*, 43–70. (Original work published 1894)

Freud, S. (1963). Introductory lectures on psycho-analysis. *SE, 15*. (Original work published 1915)

Freud, S. (1964). New introductory lectures on psycho-analysis. *SE, 22*, 3–184. (Original work published 1933)

Freud, S. (1966a). Extracts from the Fliess papers (1892-1899). *SE, 1*.

Freud, S. (1966b). Project for a scientific psychology. *SE, 1*. (Original work published 1895)

Friedman, A. (1922). Über die Krümmung des Raumes [On the curvature of space]. *Zeitschrift fur Physik, 10*, 377–386.

Fromm, E. (1973). *The anatomy of human destructiveness*.New York, NY: Holt, Rinehart & Winston.

Gaddini, E. (1972). Aggression and the pleasure principle: Towards a psychoanalytic theory of aggression. *The International Journal of Psychoanalysis, 53*, 191–197.

Gesell, A., Halverson, H. M., Thompson, H., Ilg, F. L., Castner, B. M., Ames, L. B., & Amatruda, C. S. (1940). *The first five years of life: A guide to the study of the preschool child*. New York, NY: Harper.

Gesell, A., Ilg, F. L., Ames, L. B., & Bullis, G. E. (1946). *The child from five to ten*. New York, NY: Harper.

Gillespie, W. H. (1971). Aggression and instinct theory. *International Journal of Psychoanalysis, 52*, 155–160.

Gore, R. (1983). The once and future universe. *National Geographic, 163*, 704–748.

Green, B. (1999). *The elegant universe: Superstrings, hidden dimensions, and the quest for the ultimate theory*. New York, NY: Norton.

Hartenstein, R. (1976). *Human anatomy and physiology*. New York, NY: Van Nostrand.

Hartmann, H. (1964). *Essays on ego psychology: Selected problems in psychoanalytical theory*. New York, NY: International Universities Press.

Hartmann, H., Kris, E., & Loewenstein, R. M. (1949).Notes on the theory of aggression.*The Psychoanalytic Study of the Child, 3*, 9–36.

Hawking, S. (1988). *A brief history of time: From the Big Bang to black holes*. New York, NY: Bantam.

Hawking, S. (2001).*The universe in a nutshell*. New York, NY: Bantam Spectra.

Heisenberg, W. (1927). Über den anschaulichen Inhalt der quanten theoretischen Kinematik und Mechanik [The actual content of quantum theoretical kinematics and mechanics]. *Zeitschrift für Physik, 43*, 172-198. doi:10.1007/BF01397280

Ilg, F. L., & Ames, L. B. (1955). *Child behavior*. New York, NY: Harper.

Jacobsen, E. (1964). *The self and the object world*. New York, NY: International Universities Press.

Klein, M. (1975).*The psycho-analysis of children*. New York, NY: Delacorte Press.

Koffka, K. (1935). *Principles of Gestalt psychology*. New York, NY: Harcourt Brace.

Leboyer, F. (1976). *Birth without violence*. New York, NY: Knopf.

Levin, J. (2003). *How the universe got its spots: Diary of a finite time in a finite space*. New York, NY: Knopf.

Lorenz, K. (1966). *On aggression*. New York, NY: Harcourt, Brace & World.

Low, B. (1920). *Psycho-analysis: A brief account of the Freudian theory*. New York, NY: Harcourt Brace.

259

Mahler, M. S. (with Furer, M.). (1968). *On human symbiosis and the vicissitudes of individuation*. New York, NY: International Universities Press.

Mowrer, O. H. (1960). *Learning theory and behavior*. New York, NY: Wiley.

Newton, I. (1962). *Mathematical principles of natural philosophy*. Berkeley: University of California Press. (Original work published 1687)

Piaget, J. (1954). *The construction of reality in the child*. New York, NY: Basic Books.

Planck, M. (1901). Über das Gesetz der Energieverteilung im Normal spektrum [On the law of distribution of energy in the normal spectrum].*Annalen der Physik, 309*, 553–563.

Rank, O. (1929). *The trauma of birth*. Mineola, NY: Dover.

Stein, M. H. (1972). Panel on Aggression: Panel discussion at 27th International Psychoanalytic Congress, Vienna, 29 July 1971. *International Journal of Psycho-Analysis, 53*, 13–19.

Stone, L. (1971). Reflections on the psychoanalytic concept of aggression. *Psychoanalytic Quarterly, 40*, 195–244.

Stott, C., & Twist, C. (1995). *Space facts*. London, England: Dorling Kindersley.

Weiss, E. (1960). *The structure and dynamics of the human mind*. New York, NY: Grune & Stratton.

Wolman, B. B. (1973). *Dictionary of behavioral science*.New York, NY: Van Nostrand Reinhold.

Wolpe, J. (1974). *The practice of behavior therapy* (2nd ed.). New York, NY: Pergamon Press.

GLOSSARY

Absolute death – Absolute death is the final state in the process of the body being reduced to inorganic atoms. The opposite of life is a retreat to a time before life existed. Triadic Drive Theory (TDT) calls this state "absolute death." Therefore, death itself is only a phase in the process of the retreat to the time before death, followed by organic decomposition, and then a breaking down into organic atoms.

Absolute unconscious (abs-unc) – Abs-unc is the nature of the operations of the unconscious. TDT proposes that our universe, solar system, planets, subatomic particle operations (and possibly strings), influence the atoms of the human brain in the first six months of fetal existence producing the fetal unconscious. TDT calls this the absolute unconscious (abs-unc). Gravity affects the abs-unc in causing it to sense or intuit that other forces of physics act upon humans. The content formed by the abs-unc influences the Drive-Differentiated Unconscious (DDU) in its creation of the manifest dream content.

Alternations – Alternations result from the interplay of construdo's and destrudo's struggle for dominance. Alternations are determined by absolute unconscious (abs-unc) influences on the DDU.

Anabolic processes – Anabolic are the processes that build up.

Anti-cathexis – Freud asserts the concept of anti-cathexis as when cathexis is withdrawn from the preconscious back into the unconscious. (see Cathexis).

Anxiety – Anxiety is a mood or a transitory emotion produced by construdo-destrudo fantasized alternations.

Autoerotic behavior – Autoerotic behavior is the result of libido being combined with self-construdo. In its beginnings, the libido drive was aimed at combining the embryo and the fetus within itself via physical contact, as in the folds of the embryo and fetus, giving physical contact against itself.

Basic mental functions –TDT theorizes that the principles of unconscious transformation to manifest dream thoughts are fundamentally caused by the operations of subatomic particles originating in the atoms of the fetal brain. TDT further proposes that the influence of these operations causes the human unconscious to operate as it does. This means that the forces of physics at the subatomic level cause unconscious functioning in humanbehavior. Therefore, TDT hypothesizes that the universal forces of physics activate and cause basic mental functions.

Big Bang – Big Bang is Stephen Hawking's theory that in the earliest instance of the universe a tightly compacted energy particle held together by the binding, uniting forces of physics exploded, beginning the universe.

Big Crunch – Big Church is another theory of Stephen Hawking, stating that the universe will eventually collapse back unto itself. TDT proposes that the forces of the universe operate similarly on the human psyche – because construdo and destrudo conflicts are seen to return to their undermost energy origins psychologically. (See Undermost energy origins).

Catabolic processes – Catabolic processes are the ones that break down. (See Anabolic processes).

Cathexis – In psychoanalysis, cathexis is defined as the process of investment of mental or emotional energy into a person, object, or ideas. Ideas show cathexis of memory traces, whereas emotions are processes that discharge energy, giving rise to feelings in an individual. In TDT, cathexis is the combining or connecting of unconscious drive energy with ideas, affects, fantasies, and the balance between the three drives into parts of personality and parts of one's body.

Condensation – As a salient feature of the psychoanalytic dream domain, "condensation" is analogous to "confinement" as a salient feature of physics.

Condensation process – Fusion results from the condensation process in which two or more dream thoughts and images become condensed into one dream image.

Construdo – This is the life drive, or the drive to preserve life. Construdo is the force for survival. It furthers building-up processes (anabolic) at all levels of our existence. It is the psychic drive underlying the uniting of stimulation of the senses from reality impressions, or imaginable impressions of what a realitycould be, into combinations or compositions that are unique or original. These combinations delight and please our senses by the way the combinations or compositions are joined and the way they provide stimulation.

Construdo combines sense impressions, ideas, affects, kinesthetic reactions, and attitudes, and connects them in some pattern – that is, constructs them into a definite entity, creating from them a unique, conceptualization. Further, construdo is a drive that aims to augmenting life; it is constructive and positive with regard to our existence in the world, in opposition to the destructive aims of destrudo. Construdo determines the forward in-time directions of the three drives (construdo, destrudo, and libido). Construdo may take individuals forward in time, to future fantasies that pass over the present conflicting event, and taking these individuals to a better space-time, where they deal with conflict successfully in fantasy. Construdo gives us a more complete psychological picture with the introduction of the anticipated, fantasized future as a determining factor of behavior. It also necessitates a science that can deal with predicting the probability of future events. Construdo drive is aroused when distinct unrelated stimuli from people or things confront us.

Construdo combining drive – Construdo combining drive is the fundamental drive goal, and the primary drive that continues human life, as well as the primary drive responsible for aperson's mental health.

Construdo drive – This drive is a self-combining, self connecting, self construction, and a self-creating drive. This drive re-influences the self from the earlier original stage beginning at the uterine origin. In accord with the second law of thermodynamics, in a closed system, drive's influence moves from construdo (which orders things into combinations, connections, and constructions) to destrudo, which ultimately breaks apart these combinations into disorder and disarray. It is the physics principle of entropy: order to disorder.

Construdo aim – Construdo holds the drives to their aims and objects. The ultimate aim of construdo is to unite individuals with their environment – to combine and construct a bond with that environ-ment or create a unique way these individuals can be connected with their environment. For example, love is a creative stage derivative of construdo. Love is a construdo creation of the psyche.

Construdo character traits – These are permanently etched personality traits that identify the person. Some examples are: overly connected individuals (those with an elevated need to combine with others) are considered to demonstrate a clinging trait. Overly constructed individuals are seen as workaholics. Overly creative individuals are deemed artistic.

Construdo and civilization – Construdo plus libido builds societies and civilizations. Both are necessary for civilization to exist. The conflict in civilization is between construdo and libido, versus destrudo.

Construdo and dreams – Construdo drive combines and connects the stuff dreams are made of, and then constructs and creates dreams. Construdo combines and connects intermediary images of dream thoughts with logical, rational concepts of their meaning and consequences in concert with the dream that moves in a parallel way to construction and creation of the manifest descriptive dream content. This results in two dreams. One is the actual dream – the latent dream content (the dream from below). The second dream is the dream that becomes the remembered dream, the manifest descriptive dream; its descriptive narrative.

Construdo ego – Healthy development favors the construdo ego because in contrast the unconscious psyche's perceptions, they are biologically maladaptive for the infant's survival. The unconscious psyche's perceptions were not maladaptive in utero, but become so once birth has occurred.

Construdo energy – The power and influence of construdo energy can be seen by the reservoir of construdo energy in the id or unconscious psyche that is almost twice as much as destrudo throughout most of a person's life, and about equal to destrudo and libido combined.

Creative construdo aim – Freud found that dream thoughts could be compacted, combined, and connected by the creative construdo aim in the symbolization of thoughts. According to TDT, this is another means of condensation, stemming from the confinement property in subatomic particle operations that begin in utero (fetal brain), and continue in the human unconscious following birth.

Construdo forward and backward in-time influence – With the construdo drive, which affects the unconscious, time moving forward is an ego reality. In the unconscious, time also moves backward, by destrudo influences. Libido joins in between the construdo's forward in-time influence and destrudo's backward in-time influence.

Construdo and psychoanalysis – Construdo's developing aim that achieve consciousness is what psychoanalysis has thought of as ego.

Construdo stages after birth – The development of construdo leads the infant to combine more readily and accurately, in reality terms, with its surroundings, which reflects the construdo drive's basic aim. The four stages are: 1) Self and object-combining (between 6 months and 1.5 years of age); 2) Connecting (from 1.5 to 2.5 years of age; this is a stage of play that happens when a child begins to connect its toys); 3) Constucting (from 2.5 to 4 years of age, wherein children begin to construct things with their toys for the first time); and, 4) Creating (between 4 and 6.5 years of age, wherein children's play becomes imaginative and creative).

Construdo, Object-connecting construdo – A stage in human development up to 1.5 years wherein the infant attempts to connect with all that has been registered in the infant's psyche. (See Construdo stages after birth).

Construdo, Object-constructing construdo – Object-constructing constru-do is a stage of development between 1.5 and 3.5 years of age, wherein sensory impressions are construdo combined and connected with verbal symbols or words and with perceptions of stimuli and objects in the surrounding environment. (See Construdo stages after birth).

Construdo, Object construdo drive – The developing object construdo drive is striving to combine with objects, events, affects, and images from the same surroundings, which cannot occur unless its perceptions of the surroundings are stable. Consequently, the reality perceived by the unconscious psyche must be repressed by the developing construdo drive or the conscious ego. The conscious and unconscious perceptual worlds will not co-exist.

Construdo, Object-creative construdo – Object-creative construdo is the stage of development from 3.5 years to 5 or 6 years of age, wherein the child develops creative or entirely original ways to put together impressions, images, perception, and memories of past experiences, as well as the impetus to construct new construc-tions out of all previous factors. (See Construdo stages).

Construdo, Self-combining and self-connecting construdo – These may be employed when individuals want to make some direct connection with people or relate to them in a positive way.

Construdo, Self-constructing construdo – Self-constructing construdo can be observed when peer pressure causes teenagers to construct and confirm an image given them by peers.

Construdo, Self-construdo (self-combining) – The developing embryo and fetus has bodily contact or physical combining only via touch with its own body. This stage precedes object-construdo. The work of "self-construdo" through its developmental stages of combining, connecting, constructing, and creating, that leads the unconscious processes into the new, conscious, world. This bridges the gap between the unconscious world of self-construdo and the post-birth world of conscious object-construdo's desires and motivations.

Construdo, Self-creative construdo – Self-creative Construdo is often seen between two people in a love affair. They will change themselves and adopt characteristics desired by the other, and they may copy characteristics found in the loved one. They create or mold character-ristics in themselves, which they didn't have until they thought it would please the loved one.

Construdo, Self-uniting construdo aim – Shortly before birth and at birth, the fetus has a sense of having created itself.

Destrudo – Destrudo is a term introduced into psychoanalysis by Edoardo Weiss, a founder of Italian psychoanalysis. TDT's analysis of destru-do starts from inside the individual, citing the operations registered in the brain and in the atoms of the brain. Destrudo is the force for death. Destrudo's ultimate aims are backward in time, to destroy what has been established in the individual's environment or mind. Destrudo take individuals backward in space-time. Destrudo determines the backward time direction of the three drives (construdo, destrudo, and libido). Destrudo is a drive energy that seeks to remove things that block and frustrate the individual from reaching desired construdo goals. The destrudo drive is aroused when construdo anticipations fail to materialize. Destrudo smashes wrong solutions out of existence, and then allows the correct creative construdo solution to come back into consciousness.

Destrudo aim – Destrudo's aim is to eliminate all energy, bring the life processes to an end, and to reduce these life processes to an inorganic state.

Destrudo arousal – Destrudo is aroused when a person's construdo expectations, plans, and fantasies are blocked by another or by some events.

Destrudo boundaries – The energy of such boundaries derives from the unconscious.

Destrudo, Break-apart self-destrudo aim – This is the most devastating and most basic self-destrudo aim. Occurring in utero, it causes the fetus to sense a need to destroy itself. It is what is returned to by the principle of "returning to undermost energy origins."

Destrudo in dreams – Destrudo takes the latent dream content backward to earlier times of conflicts. In dreams, backward in-time energy is supplied by destrudo.

Destrudo drive – Destrudo drive has its origin in physics. Annihilations of subatomic particles and the emerging of new particles is one origin of the destrudo drive in human beings.

Destrudo effect – This effect is accomplished by the differentiating reactions of the destructive drive's energies. The two psychic energies (construdo and destrudo) are different and antagonistic; they don't fit together, and they have different directions in time.

Destrudo energy – Destrudo energies of the archaic unconscious combine with musculature actions in the archaic psyche to form its destrudo portion. Destrudo energy must be released and expressed to acquire equilibrium for good mental health.

Destrudo's five developmental stages after birth – Destrudo's five developmental stages are: 1) Thrasher, a transitional stage from self to object-destrudo; 2) Pincher; 3) Toppler; 4) Striker; and 5) Smasher.

Destrudo, Object-destrudo – Arousal occurs when one's construdo fantasies are not met in reality, and when one experiences the blocked construdo expectations. Construdo drive frustrations arouse the destrudo drive.

Destrudo, Object-smasher destrudo and self-smasher destrudo – This type of destrudo is destroying an object so that it can never exist again. An example of self-smasher destrudo is a suicider who jumps off a building, and thus smashes the body to death.

Destrudo, in utero – This has two stages: 1) Break-apart self-destrudo; and, 2) Defective part causing self-destrudo. Destrudo has become combined with the developing musculature by the construdo operant that needs to combine things. TDT continues Freud's concept that the musculature is the executor or destrudo.

Destrudo, Pincher-object destrudo – Pincher-object destrudo is a stage of destrudo occurring between 6 months and 1.5 years, when baby develops a prehensile grasp, and when it can hold objects between its thumb and forefinger. It is occurring simultaneously with stage at which the infant's eyes can focus clearly upon objects within its environment. This gives the infant a sense of mastery and control over objects. When destrudo is aroused, the infant seeks to control the object with the aim of getting it into a vice-like grip. The infant may also grip the object with its gums. The jaw muscles and/or thumb and forefinger muscles are the executors of pincher-object destrudo. The aim is to catch an object or person with an attack from two opposite directions.

Destrudo, Self-toppler destrudo – Self-toppler destrudo will move the ideation of the original drive-aim component backward in time, before it existed in the Drive Differentiated Unconscious (DDU).

Destrudo, Smasher-object destrudo – Smasher-object destrudo is the stage in childhood development, when the child is about 5 to 6.5 years old. It is the fourth and final phase of destrudo energy. It occurs when maturation of the musculature has advanced so far that it can position an object to be destroyed by smashing it to pieces.

Destrudo, Striker-object destrudo – Striker-object destrudo is the stage in childhood development (2.5 years to 5 years of age) wherein a child wants to hurt individuals who have hurt him or caused him pain. The child verbally expresses hurts, frustrations, and resentments caused by others, and he hits or kicks the offender. Striking the object solves the child's problem. This is a learned retaliatory response that occurs when the child perceives others to have inflicted hurt or pain first. It occurs when the child begins to retaliate with destrudo-type responses. It is executed in a sudden striking attack when and where the target individual is undefended.

Destrudo, Toppler-object destrudo or Toppler-self destrudo aim – Toppler-object destrudo or Toppler-self destrudo aim is aroused in childhood if the child has been made to feel bad. In such a circumstance, the child (up to age 2.5 years old) wants to change or overturn the relationship with such people by moving them to some other place in relationship to itself. The object is "toppled" from a set position, and then pushed or pulled into a desired new position in relation to the self.

Derivative energy interactions – These are interactions that produce the emotions over those drive energy interactions that produce moods.

Dreams – In TDT, these are understood as dreams that are the product of construdo workings. The construdo drive combines and connects the stuff of which dreams are made – the actual construction of a dream. Thus, TDT ascribes to construdo the psychic energy that assembles all dreams.

Drive-aim component – When a reversal of a drive-aim component is involved because of superego self-condemning factors causing a reversal of the aim by the Drive Differentiated Unconscious (DDU) forces, but the individual senses no special talents or means of expression for solving the drive dilemma through sublimation, then a reaction formation against the original drive-aim component is likely to occur. Thus, the individual displays expression of the opposite of the drive-aim component. The universal alternating influence causes this drive alternation of reversal.

Drive or Drive derivative alternations – These cause psychic energy iteration of friction between these two energies, as does their space-time direction. Drive-aims are forward, while others are backward in space-time. The friction between the drive alternations and drive derivative alternations produces human moods such as elation, depression, and happiness.

269

Drive derivative interactions – Human emotions are caused also by drive derivative interactions and by the friction between two energies – derivative interactions and drive energy interactions. In the derivative energy interaction producing the emotions over drive energy inter-actions that produce moods, there is a refinement of energy or rarefied energy related to the emotions.

Drive Differentiated Unconscious (DDU) – The drives (construdo, destru-do, and libido) differentiate into the drive-differentiated uncon-scious (DDU). The drives emerge into the DDU in the following order: first, Construdo; next, destrudo; and finally, libido. The interaction in the DDU between destrudo and construdo, and in subsequent space-time, parallels the changes in time directions that originate in the absolute unconscious (abs-unc). The DDU evolves from 6 to 9 months in utero, and in the first 6 months after birth. The DDU shows that drives operate more and more independently of each other, seeking differentiated goals. This unconscious level is distinguished by the unconscious of the fetal brain before six months in utero. This independent functioning of the drives is seen in the abs-unc. It occurs before all three drives' energies began to disconnect and separate from each other, thereby differentiating as in the DDU. The unconscious component of the drive ideal is a resultant construct (when some mediating psychic proposition is needed to give closure and reconciliation to the conflicting drive influences from physics and biology, which are paradoxical), and it is created in the DDU. Construdo disconnects the drives and their different energies during the third trimester's fetal psyche, which than becomes the DDU.

Drive energy – This is the energy of construdo, destrudo, or libido. A drive'senergy should never be impeded or blocked from its arousing stimulus. The drive's energies should be focused on the situation at hand and not be mired in the unconscious past or in imagined future unresolved drive conflicts. Drive energy from the abs-unc and the DDU are interchangeable, allowing substitutions of one drive energy for the other. Drive energies always go back in time to the undermost energy origins of the drive involved. This return helps explain the vicissitudes of psychic drive expenditures of energy and the final direction such drive energies follow. A drive will not discharge unless it has an exit and a return pathway for replenishing itself in its unconscious origins. This is the "replenishment principle" of TDT. In TDT, when a drive's energy returns from consciousness back into the unconscious, its aim can be reversed by the unconscious. (See Absolute Unconscious [abs-unc]).

Drive fusions – This is the result of the energy of construdo used to unite other drives. For example, the fusion of libido and destrudo involves the energy of construdo uniting them.

Drive-ideal – This is the disconnected part of Construdo, with its logical thoughts, feeling, and sensations becoming reconnected in the psyche when these create a new mental organization. The drive-ideal is a focus of drive energy between the DDU level and consciousness. (See Drive-Differentiated Unconscious, DDU).

Drive response – The closer the drive response is (which corresponds to/ correlates to the conscious psyche's reality), the healthier (or more "normal") this person's response will be.

Drive regression – This is the result of the "principle of returning toward undermost energy origins."

Ego – Construdo's developing aims that achieve consciousness are what psychoanalysis has thought of as ego.

Electromagnetic force – Electromagnetic force holds together atoms and living molecules, cells and connective tissue, organs, organ systems, and skin, binding our entire bodies together. It also holds everything else on our planet together. The molecules of all animate and inanimate objects are held together by electromagnetism, as are parts of atoms that comprise them.

Exclusion principle – The exclusion principle of modern physics states that two subatomic particles of identical spin – half spin particles – cannot (within the limits of Heisenberg's uncertainty principle) have the same position and velocity.

Fetal unconscious – TDT proposes that our universe, solar system, planet, and subatomic particle operations (and possibly strings) influence the atoms of the human brain in the first six months of fetal existence, producing the fetal unconscious. TDT calls this the absolute unconscious (abs-unc).

Four forces of gravity – Four forces of gravity are: gravity; the electromagnetic force; weak nuclear force; and strong nuclear force.

Freud's Dual-Drive theory – Libido and aggressive (destructive) drive theory.

Future – The future lies in the space-time direction of construdo aims as demonstrated by human wishes, daydreams, fantasies, and future plans.

Gravity – Gravity holds our bodies, alive or dead, to the Earth. It is the force holding our solar system, our galaxy, our entire universe together. Gravity is one of the forces of physics that creates the construdo drive.

Human emotions – Emotions are elicited when one or more of the three basic drives is/are involved in an interaction with each other or with the environment.

Libido – Libido is the biological drive in the human psyche that directs human beings to come together, sensually and sexually, for the hidden, not consciously considered purpose of recreating the human species. Libido reproduces or recreates life. Libido may attach itself by construdo either forward in time (as in a sexual drive towards another) or backward in time (as in a dream that reworks sexually repressed wishes and finds release of the blocked wish). Libido, while it has its own terms and principles, is also pulled into dynamic interactions with construdo and destrudo by the construdo drive. Libido occupies a middle position between destrudo and construdo aims. (See: Construdo forward and backward in-time influence).

Optimal Existence – For any person or all of humankind, Optimal Existence is one in which, mentally, there is a clear dominance of construdo-type energies (joining, uniting, building, and creating).

Psychic drives – These are drives (construdo and destrudo) derived from physics. They define boundaries of human behaviors; libido operates between them. The drives (construdo and destrudo) move in opposite direction under conditions of forward (in consciousness) and backward (in unconscious). Thus we see both, time moving backward in a dream and time moving forward in consciousness.

Psychic energy – TDT's position is that substituting construdo energy for destrudo leads to a mentally healthy release of built-up psychic energy.

Psychopathology – Psychopathology can arise when one or more of the drive-impulses is/are distorted or transformed as they enter the ego from the id in space-time. In other words, psychopathology is likely when there are differences in the dimension of space-time between the realities of the universe and the realities in consciousness – such as in an individual's ego, for one, two, or three of the drives (destrudo, construdo, or libido).

Reaction formation – Reaction formation is a displacement from self-destrudo to self-construdo. An example is when a drug addict becomes a drug rehabilitation counselor who helps reclaim people from efforts to destroy their own lives.

Reality principle – The reality principle is a result of construdo's aim (second stage) to form connections with the surroundings. This was not explained by Freud, but rather described. Construdo's overriding aim to connect the individual with reality was called by Freud "the reality principle." When an infant's sensorium can well perceive its environment, the forces of physics drive it to construdo connecting with the environment. Now the infant can focus its eyes on things around it. Such perception seeks constancy and predictability from construdo's overriding aim to connect the individual with reality. The attainment of the reality principle separates construdo from libido and destrudo, in the id – in the unconscious.

Regression – Regression can be observed when a person's consciousness is dominated by past, earlier in-time unconscious construdo fantasy determinations or dream determinations of what was reality. Regression is a return to the first grouping of the three undifferentiated drives of the unconscious, a psychotic condition: to pure abs-unc (absolute-unconscious), or to the advanced DDU, as well as to the alternations between the two (still a psychotic condition).

Repetition compulsion – Repetition compulsion is what Freud called a compulsion to unconsciously repeat the same failing solution to the earlier appearance of the same unsolved problem. The problem is never solved because the old solution is unconsciously reapplied.

This compulsion demonstrates the principle of returning to undermost energy origins.

Replenishment principle – The replenishment principle of psychic energy is akin to the conservation of energy principle of pre-modern physics, which states that energy cannot be created or destroyed. A drive will not discharge unless it has an exit and a return pathway for replenishing itself in its unconscious origins. This is the "replenishment principle" of TDT. The replenishment principle indicates that the energies of two basic drives can be substituted one for the other in the Differentiated-Drive Unconscious (DDU).

Replenishment of drive energy principle – Destrudo aims in the DDU can be transformed into a righteous drive-ideal cause. The DDU can reverse destrudo aims into a positive conscious-satisfying bonding. This reversal substitution of energy is due to the replenishment principle of drive energy.

Repression of drive aims – Here, construdo's normal forward direction in space-time is reversed by the backward energies of destrudo from the DDU. Once set in this direction by the pull of the reversing energies from the abs-unc, complete repression of the drive-aims results.

Psychosis – When an individual's psychic regressed drive energies begin to alternate between the DDU and the absolute-unconscious (abs-unc), psychosis results. The greater the frequency and intensity of these alternations are, the more severe the psychosis.

Self-combining aim – Sublimation of the self-combining aim after birth – an ambivalent trait – is seen in one's ability to compromise two opposing positions.

Self fault-finding trait – Self fault-finding trait is seen when individuals strive to intensely develop whatever talents they find in themselves.

Space-time – "Space" relates to events of incidents in one's life experiences and where such events occurred. "Time" relates to when these events occurred. As space-time relates to TDT, "space" refers to a psychological event at a given place in one's life; "time" refers to when the event occurred. The number of space-time coordinates refers to the number of coordinate points necessary to delimit a beginning event in the psychological space-time, and its subsequent

274

influences on succeeding space-time events leading up to the current and future space-time events. Physicist Stephen Hawking said that "time is not completely separate from and independent of space, but is combined with it to form an event called space-time." The conflict with time and its direction forward or backward, as well as the thought content (space) derived from a drive's energy or the space it occupies, in terms of what we think our lives are about, is a unit of physics, or a new way of looking at human events referred to by physicists as space-time. Modern physics asserts a new dimension of space-time by considering the universe and its forces, as well as the forces within the atom. Both, universal forces and the forces within the atom influence what becomes the human unconscious.

Space-time dimensions – "Normality" implies the fewest transformations of distortions in space-time dimensions when drive impulses enter the conscious ego's reality. In contrast, "psychosis" implies transformed, distorted, and confused space-time factors in all three drives. Therefore, the space-time dimension is an overarching conception applying to the reality of the universe beyond us, to us on Earth, and to the subatomic operations within our brains.

Strings – TDT speculates that strings are an alternating energy state along with the matter state of particles.

String Theory – A string is the smallest, non-divisible constituent of a subatomic particle. It is theorized that strings show a constant oscillation of energy. Strings are thought to underlie all matter and interactions, thus explaining gravity and other forces, as well as particles in everything and everywhere. In this respect, String Theory has been called "the theory of everything."

Strong nuclear force – Strong nuclear force holds together the subatomic protons and neutrons in the atomic nuclei of our bodies, as well as in the other nuclei of atoms on our planet, and in our universe.

Sublimation – Sublimation is the displacement of the drive's component-aims. Sublimations are reversal transformations turning shame, disgust, and oral self-condemnation into aims that will gain social approval and approbation. Such reversals are the work of the absolute-unconscious (abs-unc) processes on the DDU. When an individual's destrudo breaks down a reality, and then construdo reconnects it with a new reality, this is a "sublimation" of the individual's break-apart destrudo aim. Here, the trait is an alternation of the aim.

Thrasher self-object stage – In infancy (birth to six months) when an infant venting, cries and screams, its musculature is essentially attacking the unknown. The baby had yet to differentiate its target as inside or outside itself. This stage is a product of self-destrudo aims that conflict with object-destrudo aims.

Time – Time is a factor creating distortion in dream content; that is, time must elapse before the second drive's entrance into the dream.

Triadic Drive's influence – This drive begins in utero, and TDT predicts that its effect produces the fetal unconscious.

Triadic Drive Theory (TDT) – TDT is a theory of how the forces of the universe create our mentality. TDT is derived from the concept that humans exist in a four-dimensional world of three space dimensions and one time dimension. TDT will show that destrudo's and construdo's various stage-aims are also involved in trait formation as found with libido. TDT asserts that human behavior and psychic ideation is driven by libido, destrudo, and construdo. According to TDT, the drives do not fuse as was long proposed by psychoanalytic theory. TDT infers that drives do not fire simultaneously at the same target, but that the drives fire sequentially.

Uncertainty Principle – Werner Heisenberg stated that the more precisely the position of a particle is determined, the less precisely its momentum can be known, and vice versa. In TDT, whether drives will be directed forward or backward in time is comprehensible by the consideration of the Uncertainty Principle. The Uncertainty Principle governs the outcome of influential factors, in terms of what will be observed as an ultimate outcome.

Unconscious – Unconscious includes the processes of the mind that occur automatically and are not available to introspection. These include thought processes, memory, affect, and motivation. Though these processes exist without conscious awareness, they impact one's behavior. In TDT, the unconscious operates as it does because the principles that govern the operation of subatomic particles in the atoms of our brains create a field of influence in the fetus's developing brain. The nature of the operations of the unconscious is stored in TDT's absolute-unconscious (abs-unc). With the construdo drive (which affects the unconscious), time moving forward is an ego reality. In the unconscious, time also moves backward by virtue of destrudo influences. Libido joins in-between the construdo's (forward) in-time influence and destrudo's (backward) in-time influence. Freud found that the unconscious was timeless in its operations, just as the operations of subatomic particles are in the atom. Spatial relations have no meaning in the unconscious or in subatomic particles. Moreover, in the unconscious, there is no logic as we know it, no logical contradictions or negations. Similarly, subatomic particles operate without logic. The unconscious is not affected by contradictions and alternatives because conscious reality is not present in the unconscious.

Unconscious psyche – The unconscious psyche is the source of destrudo, construdo, and libido, as well as at the source of human creativity. The energy of the unconscious psyche causes the issue of striving to fulfill the human need to combine, connect, construct, and create a union with things in the world following birth. In the unconscious psyche, construdo and destrudo come into existence as unified, organized, and opposing psychic drive-forces. Freud asserted that an instinct or drive resulted from a constant influence or pressure on the unconscious psyche. The unconscious psyche's perceptions were not maladaptive in utero, but became so once birth occurred. The unconscious perceptions are not a part of the world into which the infant has been delivered, but according to TDT, these unconscious perceptions belong to the world of subatomic particle operations that account for the electrical charged brain field that exists in utero.

Undifferentiated chaos – Jean Piaget first noted that the infant rejects the constantly changing perceptions of its unconscious.

Underlying energy origins – When individuals become overly invested in any construdo stage-aim, the focus to return to underlying energy origins can lead to the formation of construdo character traits.

Undermost energy origins – A principle that indicates that exhibited mental energies in a present mental event ultimately return to that place in space-time, from which the drive energy interactions originated – to earlier states of existence before there was life. This is the ultimate aim in the psyche of destrudo against the self. Drive regression is the result of the "principle of returning toward undermost energy origins." Animals that migrate to the locality where they were born all show tendencies to return toward the undermost energy origins of their construdo.The principle of return to "undermost energy origins" helps explain the vicissitudes of psychic drive expenditures of energy and the final direction such drive energies follow.

Universal alternating force – These alternations may be from one drive to another, or to two others. Such transformations of destrudo aims into construdo aims, or into construdo and libido aims, occur back and forth in the DDU caused by alternating energies from the abs-unc, which in turn, influences the DDU.

Uterine energy – Uterine energy of self-construdo is the beginning of all sense of self in humans.

www.ingramcontent.com/pod-product-compliance
Lightning Source LLC
Chambersburg PA
CBHW070842300326
41935CB00039B/1364